Applications of Augmented Reality - Current State of the Art

Edited by Pierre Boulanger

Published in London, United Kingdom

Applications of Augmented Reality - Current State of the Art
http://dx.doi.org/10.5772/intechopen.111231
Edited by Pierre Boulanger

Contributors
Angelo Compierchio, Phillip Tretten, Prasanna Illankoon, Peter Kudry, Michael Cohen, Pierre Boulanger, Pratik Pradhan, Mohsen Rostami, Jafer Kamoonpuri, Joon Chung, Sara Rankohi, Mahsa Rezvani, Lloyd Waugh, Zhen Lei, Matevž Dolenc, Igor Agbossou, Juan David Rueda Martínez, Gonzalo Andrés Moreno Olano, Sebastian Roa-Prada, Hernán Darío Mantilla Hernández, Nestor Fernando Saavedra Trujillo, Quentin Neuville, Thierry Scheerlinck, Johnny Duerinck, Klaudia Proniewska, Damian Dolega-Dolegowski, Radek Kolecki, Magdalena Osial, Agnieszka Pregowska, Călin Corciovă, Robert Fuior, Andra Băeşu, Cătălina Luca

First published in London, United Kingdom, 2024 by IntechOpen
IntechOpen is the global imprint of INTECHOPEN LIMITED, registered in England and Wales, registration number: 11086078, 5 Princes Gate Court, London, SW7 2QJ, United Kingdom

British Library Cataloguing-in-Publication Data
A catalogue record for this book is available from the British Library

Additional hard and PDF copies can be obtained from orders@intechopen.com

Applications of Augmented Reality - Current State of the Art
Edited by Pierre Boulanger
p. cm.
Print ISBN 978-1-83769-335-1
Online ISBN 978-1-83769-334-4
eBook (PDF) ISBN 978-1-83769-336-8

For EU product safety concerns:
IN TECH d.o.o., Prolaz Marije Krucifikse Kozulić 3, 51000 Rijeka, Croatia,
info@intechopen.com or visit our website at intechopen.com.

Meet the editor

Dr. Pierre Boulanger has more than 35 years of experience in medical imaging, rapid product development, and the applications of virtual and augmented reality systems to medicine and industrial manufacturing. He is a professor in the Department of Computing Science and the Department of Radiology and Diagnostic Imaging, University of Alberta, Canada. He is currently the director of the Advanced Human-Computer Interface Laboratory (AHCI) and the scientific director of the SERVIER Virtual Cardiac Centre. His main research topics include the development of new techniques for telemedicine, patient-specific modeling in surgery, and the application of augmented reality technologies to surgical training, simulation, and image-guided procedures. His work, including more than 400 scientific papers and collaboration with more than 10 universities, research labs, and industrial companies across the world, has contributed to his gaining international recognition in the field of augmented reality.

Contents

Preface

Augmented reality (AR) is a technique that combines real-world experience with computer-generated sensory outputs to assist humans in accomplishing complicated tasks. The requirement for AR lies in the need for real-time interaction and precise 3D registration of both the virtual augmentation and the real world. If it is achieved, a distinct link between physical reality and digital information is established that can be used in multiple applications. To be successful, AR systems need to integrate technologies like electro-optical systems, real-time computer vision, high-performance computer graphics, cutting-edge human–computer interaction, and complex software tools to integrate digital information into reality. This book presents various AR applications, including real-time information display, applications in the construction industry and architecture, and finally medical applications. Each chapter describes the application of AR in these areas and describes the current state of the art. My objective in editing this book was to achieve a balance between principles and practice, illustrating applications from technical, methodological, and user perspectives.

Pierre Boulanger
Department of Computing Science,
University of Alberta,
Edmonton, Canada

Section 1

Augmented Reality for Real-Time Information Display

Chapter 1

Augment-Me: An Approach for Enhancing Pilot's Helmet-Mounted Display Visualization for Tactical Combat Effectiveness and Survivability

Angelo Compierchio, Phillip Tretten and Prasanna Illankoon

Abstract

A learning framework for combining state-of-the-art augmented reality (AR) technologies and artificial intelligence (AI) for helmet-mounted display applications in combat aviation has been proposed to explore perceptual and cognitive performance factors and their influence on mission needs. The analysis originated through examining helmet-mounted display (HMD) design features and their configurations for tactical situational awareness (SA). In accomplishing this goal, the relationship between the pilot visual search and recent advancements in AI have been gauged as a background source to unlock pilot's uncued visual search limit. In this context, the Augment-Me framework is introduced with the ability to view and organize SA information in a predictive way. The provisioning of AI-augmented fixation maps could effectively outperform current AR-HMD capabilities, facilitating human decision while pursuing the detection and compensation of the mechanisms of human error.

Keywords: augmented reality, helmet-mounted display, situational awareness, artificial intelligence, visual search, tactical readiness, fixation

1. Introduction

The dynamic evolution of global challenges has endured many Air Forces in seeking more innovative and financially optimized solutions to enhance pilots' performance and learning within a wider combat envelope. Wearable solutions have led researchers to rely on the use of advances in technology to examine the tactical information needs of pilots. This explanatory step has led to the development of state-of-the-art devices for processing, transmitting, and displaying images onto the pilot's visor. Revolutionary interface designs have extended the helmet line-of-sight from virtual images enabling the pilot to see a target through the aircraft to an uninterrupted display of navigation imagery both day and night. These developments coined with the Distributed Aperture System (DAS) capable of warning the pilot of incoming threats have made the F-35 stealth aircraft as the most capable SA aircraft [1]. Such an unrivaled capability accelerates decision-making with unmatched data sharing networked for warfare operational vision [2].

Operational vision boundaries have been confronted with the paradigm of linking computational-oriented knowledge and naturalistic recognition of decision-making. This recognition has been evaluated from the perspective to integrate event-driven and goal-driven outcomes. VR augment cockpit information, however, for effective airpower intelligent support is needed as the human central vision is limited to roughly two degrees wide [3]. When the senses can provide accurate and depth information, the inherent perception and action connection represents an opportunity for fast cognitive processing of maneuver-related information. Seeing is fixating, humans need to fix the gaze on a single spot to inspect the minute details of the world [4]. However, often some stimuli convey distorted images or sounds that cannot be easily perceived, as are biases to sensory processing. This limitation can be overcome if a stimulus has been primed with other stimuli. In this context, the Augment-ME framework explores this discrimination with AI augmented content through a *"virtual pilot"* to enhance threat detection when engaged in combat conditions. There is an ongoing trend in implementing AI in the combat domain on either side of the Atlantic. Recent reports indicate that Su-57 Fighters are to Get "Smarter" with AI-enabled sensor fusion and data processing. A report stated that: *"intelligent support of the leading and slave pair of fighters in the process of conducting long-range air combat with a pair of enemy fighters"* [5].

This definition emphasizes two aims a "pair of fighters" as a two-on-two scenario and "long rang" as beyond visual range combat.

Against potential opponents, the tasks of the virtual pilot are to uncover potential enemy threats in a reasonable amount of time with fixations through an eye-tracking system. This measurement concerns recording eye movements of the pilot and is examined by eye and pupil metrics [3, 6–10]. An AI algorithm can be developed to estimate the position and duration of fixations as shown in **Figure 1**.

The main purpose of a pilot is to rely on the eye as the primary sensor to retain critical SA elements and establish cognitive dominance, as defined: "Conscious awareness of actions within two mutually embedded four-dimensional envelopes" [11], this relationship is portrayed in **Figure 2**.

Therefore, the evolving pilot's mental model relative to the time, position and direction of the aircraft is represented through the inner and outer envelope of the SA construct. The role of the inner envelope refers to a pilot positioned through Boyd's Observe/Orient/Decide/Act (OODA) loop in an unaided sensory space, while the outer envelope refers to the pilot receiving assistive AI information. This task is implicit as the pilot can bypass the explicit part of the loop almost simultaneously. Moreover, a pilot engaged in air-to-air combat in a challenging SA environment poses

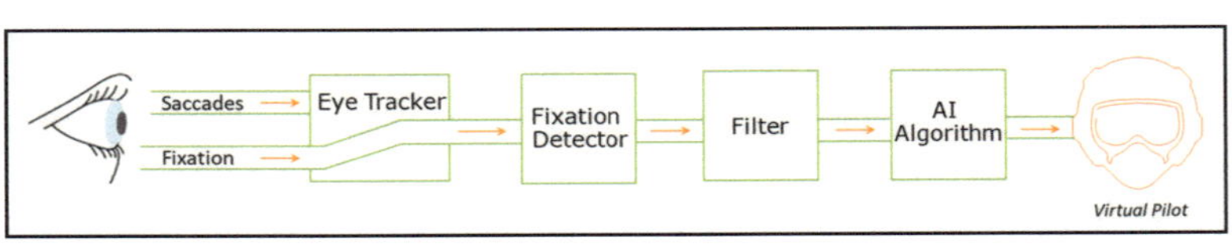

Figure 1.
The virtual pilot.

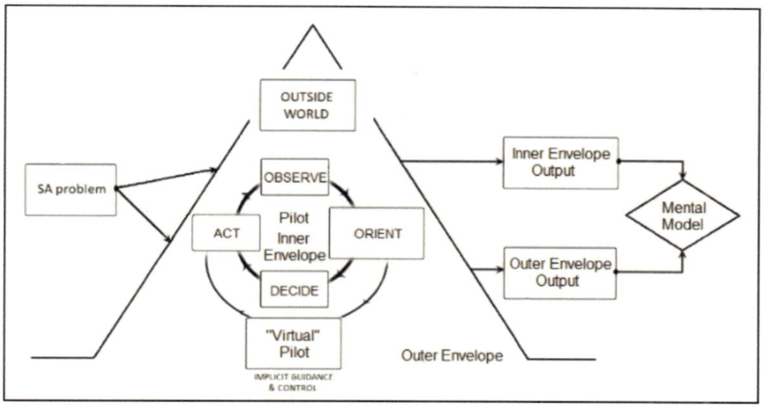

Figure 2.
Positioning the virtual pilot in a mixed real-virtual environment, (adapted from [12]).

Figure 3.
Pilots' optical flow (source adapted from public domain).

higher demands on locomotion and stimuli response. These important aspects refer to the optical flow of pilots' view, in relation to the eye sensitivity to motion and the field of regard (FoR) which is the local motion within the total area that can be perceived [13]. A frame from a MotoGP event in **Figure 3** shows this effect in action.

The optical flow is presented by vectors representing motion velocity on the track. The scenario of moving forward presents arrowed objects that although the rider accelerates and decelerates apart, they converge toward each other. This sequence makes the fixation on each of the paired objects relatively easier to detect. In principle, this paired category yields the perceptual relationship of object motions in relation to the background.

This approach is intended to uncover fixation time-based maps and to maximize the chances of detecting a target through additional spacetime, as well as capture focused attention as expressed in SA space terms by the U.S. Department of Defense (DOD) dictionary: "*The requisite foundational, current, and predictive knowledge and characterization of space objects and the operational environment upon which space operations depend*" [14]. The visual engagement of a target remains the main focus in aerial victories, but humans are mostly capable of making only two to three fixations per second. In this study, an AI-augmented approach has been proposed to leverage the effects of falling acuity when the direction of the eye is stirring from one fixation point to another.

2. Evolution of AR-HMDs

The pace of technological advances embedded in combat flight helmets has been relentless. Just over a decade after the Wright Brother first flight in 1903, Albert Bacon Pratt patented an integrated gun helmet, this conceptual helmet-mounted sight (HMS) was worn with a platform for weapon delivery and control [15]. Undoubtedly, the combined integrated approach was also awarded US and UK patents for an "Integrated helmet mounted aiming and weapon delivery system". This was the first known conception of combat flight helmets as an integrated system. From the mid-1930 till 1940 the Type B Flying helmet was developed with improved and new design features such as oxygen masks, radio earphones and removable goggles. The leather helmet was later replaced by a hard helmet to add further protection to pilots when forced to eject at higher speed, while the goggle was replaced with a tinted visor to protect the eyesight from harsh sunlight. In the mid 1950s Gordon Nash reportedly researched new methods to equip pilots with additional information, further envisaging the pilot's constant need for operational and tactical information for combat survival. During WWII the advent of the radar changed warfare development on a wider scale, greatly influencing the dominion of airspace and the role of fighters.

This revolutionary development was also tailored with a fire control system to assist pilots to hit a target by supplying both the range and the direction of the target. This device known as a fire-control radar designed to improve mission performance remains a proven capability these days. A fighter aircraft equipped with a fire-control radar would have tactical superiority from being practically invisible to the naked eye, therefore, a pilot to prevail had to have knowledge of flight navigation and weapon information. The thoughtfulness of presenting flight referential conditions to the pilot led to the development of the Head-up Display (HUD), a fixed, transparent and localized display that presented flight data without obstructing the user's view. The HUD allowed the pilot to keep focus on the image and attention on the forward airspace, unlike, earlier Head-Down Displays (HDD) that presented aircraft status information and required the pilot to look inside the cockpit for updates. One aircraft equipped with a real true HUD was the Havilland Mosquito, allowing a pilot to view radar information and an artificial horizon directly in front of flight controls. The addition of the horizon was instrumental for flying as without the pilot could not tell the aircraft's position and lose the sense of balance or position [16]. Most obviously the HUD also presented some drawbacks to pilots acquiring and maintaining the SA. The pilot could only receive collimated imagery as of target information only on a small-forward looking area of the aircraft, then a viewing device was needed to always provide information directly in front of the pilot's eyes. The solution originated in the 1960s with the design and development of the HMD as it provided the pilot unobstructed visual field and unrestricted head movement with heads-up capability. In physiological terms, this translates to better eye relief and lower weight.

2.1 Technological trends

The HMD enhanced the fighter's offensive capability by allowing pilots to choose a target only by looking at a guileless perceived threat. A common classification for HMDs refers to the type of presentation of the sensory image and symbology, which include monocular which refers to one image to one eye, biocular or the same image to both eyes, or binocular or two different images to each eye.

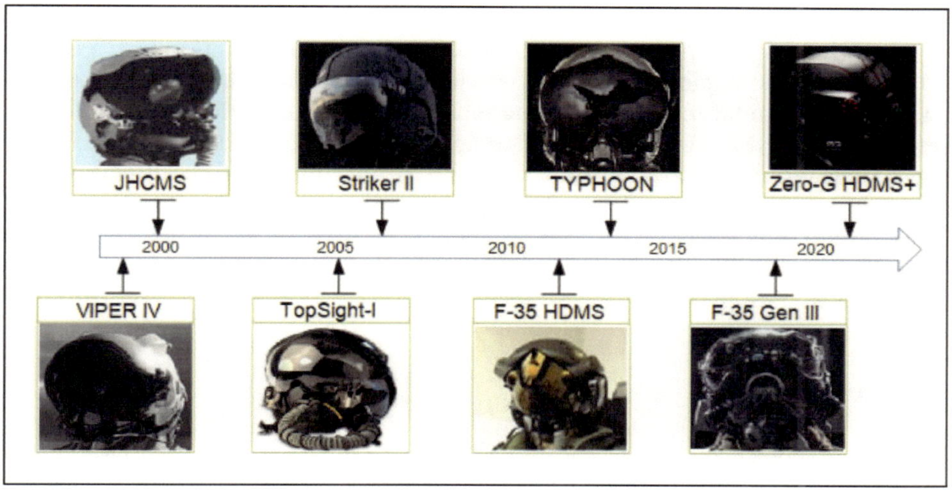

Figure 4.
Summary of HMD programs (helmet source: Public domain).

The Israeli DASH became the first known HMD that measured the pilot's line of sight (LOS), slaved missile and sensory data to the target location while integrating all aircraft operational modes while the pilot had Hands On Throttle And Stick (HOTAS) controls. The DASH formed the basis for the Joint Helmet Mounted Cueing System (JHMCS) program. JHMCS technology was developed through a joint venture established by Elbit Systems, Vision Systems International (VSI) and Kaiser Electronics. The JHMCS equipped the F-16, F/A-18, F-15 and F-22, it gave the pilot advanced helmet sight despite reduced Air-To-Ground (A/G) accuracy performance due to known timing drift issues related to the cockpit magnetic fields. An overview of fixed-wing HMD development programs is shown in **Figure 4**.

Earlier HMDs developed as the JHMCS, the TopSight-I and the VIPER-IV HMDs employed a monocular visor-projected display and retained sensory technology to determine the pilot's line-of-sight. This trend, thereafter, from TYPHOON to the F-35 steadily shifted toward binocular visor-projected display reflecting the potential benefits of this classification as decluttering information, reducing perceived workload, improved capturing attention, search times and reducing response times [17]. The projection of AR-style symbology on the F-35 HMDS evolved from simple flight data to full motion video, image insertion and color symbology. The Helmet Mounted Display System (HDMS) used the same JHMCS symbology and was the first helmet to incorporate HUD functionality.

Ironed challenges of the HDMS benefited the roll-out of the F-35 Gen III helmet that included the resolution of blooming effects especially during night-time, it was discovered the employed liquid-crystal display (LCD) active-matrix (AM) displayed on the HMD visor a green haze from the backlight. This finding made it difficult to locate targets in dark conditions. The solution focused on replacing LCD with organic light-emitting diode (OLED) displays which eliminated the bleed-through green glow. A change was also made to the helmet tracker to meet helmet line-of-sight accuracy strafing requirements [18]. The Gen III helmet became fully operational with two ocular cameras situated above the helmet's display visor. The cameras are manually aligned to the pilot's interpupillary distance (IPD) settings for each eye. This task is set to optimize the acuity of each pilot's vision as well as establish the proper fit of each helmet. The next generation of binocular HMDs (Zero-G HDMS+) is set to empower the 4th, 5th and 6th with improved optical acuity, SA and information management.

However, technological advances have added additional emphasis to these trends on the physical limitations of the eye as a sensor. Despite innovative technologies, there is ongoing HMD research on F-35 topics influencing human performance: green glow, double vision, monocular symbology and canopy interaction [19]. These aspects cover the basic visual search demands that can be commensurate with an HMD cueing system, however, eye-tracking technology is currently not available on aircraft HMDs. This technology would enable to investigate pilot's scan patterns necessary for high maneuverability.

3. The augmented SA

Air Forces engage reconnaissance squadrons to maintain tactical readiness and to conduct close air support (CAS) operations. CAS is denoted as *"air action ... against hostile targets that are in close proximity to friendly forces... (requiring). .. detailed integration of each air mission with the fire and movement of those forces"* [20].

The ability to sustain CAS capability requires eye-blink situational awareness with advanced integrated avionics. The deployment of the F-35 stealth aircraft to provide CAS requires experienced pilots capable to converse decision-making directly on the edge. This need was already perused by earlier known pilots such as Oswald Boelcke, the central theme in combat tactics laid on the basic concepts of analyzing the enemy's position, executing counteroffensive maneuvers and making continuous decisions. As the recognition of decision-making depends on the knowledge structure directly recruited through time-based tactical appreciations, the complex percepts involved were codified by the OODA loop. This reveling interaction evokes Gibson's work in stating that the pilot's sensory system abided the percept construct directly from the environment, such a thought led to establishing the Theory of Direct Perception (TDP) [21]. This theory introduced the concept of affordances related to the interaction between the entire world and the pilot as if the sensory system would be resonating with environmental features.

The effects of this interpretation were quoted as: "*When the senses are considered as perceptual systems, all theories of perception become at one stroke unnecessary. It is no longer a question of how the mind operates on the deliverances of sense or how past experience can organize the data, or even how the brain can process the inputs of the nerves, but simply how information is picked up*" [22]. The picking of maneuver task-relevant information is key to minimizing workload and increasing SA, while the symbology flow afforded to pilots on display concepts as the HUD provided the opportunity to intercede constraint performance on cognitive tasks [23, 24]. Essentially, the HUD was a genuine precursor to AR breakthroughs in the aviation domain. The two AR words as a phrase were coined by Tom Caudell and David Mizell [25]. The overlay of virtual guidance symbology on the pilot's out-the-window view provided comprehensive situational awareness (SA) improvement enabling the pilot to filter a larger amount of information into relevant SA inputs. The HUD essentially provided the pilot with the means to remain "head-up" despite instrumental meteorological conditions (IMC). Furthermore, the combination of real and augmented cues provided several aids to the pilot, including increased flight safety related to improved warning indications, reduced pilot workload based on flight completeness of information, increased flight precision and direct trajectory visualization. The Augment-Me framework follows

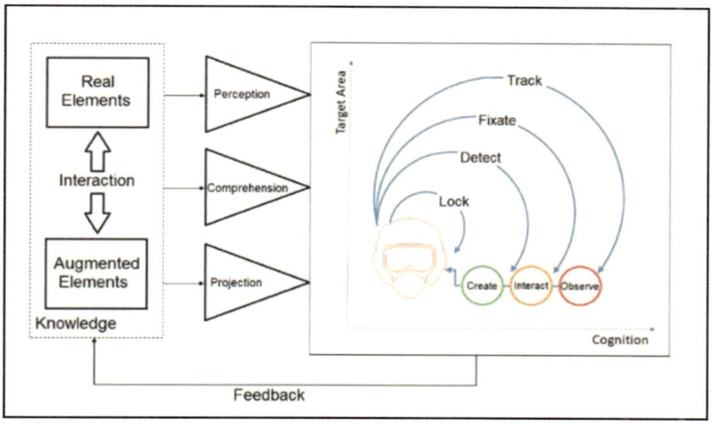

Figure 5.
SA interface model.

the reflection of visual and control augmentation combination, their influences on cognitive thinking and endows the role of the SA. This ability respond to different SA levels as "Level (1) the pilot's accurate perception of elements information within the cockpit, the aircraft and the external world, Level (2) the comprehension of the significance of those elements, to form a knowledge pattern with a holistic temporal picture of the external world and Level (3) the ability of the pilot to project future actions through the knowledge acquired from the projection of the elements status in the near future" [26]. This description is illustrated in **Figure 5**.

4. Augmented visual search

The prospect of superimposing and projecting on the HMD visor virtual information with visual cues empowered tactical maneuvering and delimited attentional capture. The first HMD that exploited AR was the Integrated Helmet And Display Sight System (IHADSS). The IHADSS was a monocular device displaying a virtual image on a single eye, this feature led to increased workload and stresses as a result of projected visual disturbance due to head movements. This layout triggers competition between both eyes affecting visual search. On the other hand, an alike condition caused by the neural adaptation of switches between the perceptual state also characterizes binocular rivalry by facilitating visibility near the point of fixation rather than the fixation itself [27, 28]. A consideration for monocular augmented reality is that there is competitive dominance between stimuli luminescence and moving or stationary stimuli. While for binocular rivalry when wearing see-through displays experiments have shown that the influence of stimuli was less unstable than in monocular displays [29]. Notably, this can cause visual fatigue as binocular bistability could incur due to the image alteration between both [30]. The dynamics of every single action starts with the eye and end with the eye and an eye reaches an object before the completion of the head motion.

As such, the distinctive ontology between pilots and combat situations is that situations are structured while pilots do not. This is remarked by the exemplified same human components of a pilot that flew the Tornado aircraft in the late 1970s and a pilot who flies the F35 today. Besides the technological advances the importance of the pilot visual search has not changed, a pilot would still have the last say, whether they are performing Active air and missile defense or Passive air and missile defense. Active defense refers to containing the enemy through kinetic and non-kinetic actions, while Passive defense includes all measures except detection and warning assets [4]. This leads to the instantaneous and continuous evolution of the pilot's mental picture invoking the dynamics of visually guided reaching [31].

4.1 Visual search measures

Visual search is made up of a series of discrete fixations [32]. Fixation refers to the visual gaze that shifts to a location for a period, between this shift saccade forms. This formation creates 3–5 saccades per second with fixations occurring for about 300 ms, while longer fixations are representative of higher cognitive processing [33]. Saccadic eye movements inhibited between visual attention patterns are relatively brief, their movement has a central function of forming fixation. Eye-tracking devices require sufficient spatial and temporal resolution and must capture in real time a wide range of eye movement measures [34–41]. Eye-trackers are non-intrusive and can be monocular or binocular and following calibration can measure eye movements of the line-of-sight direction along both the x-axis range and the y-axis range. It is worth noting that the line-of-sight is defined from the direction of the eye gaze and the head. On aircraft, video images generated from sensors mounted on the aircraft or the HMD are directly correlated with the eye-of-sight. Sophisticated HMDs have been built with incorporated eye and head tracking systems. Generally, most eye-tracking devices have video-based capability and are categorized as mobile and remote tracking systems. The former refers to head-mounted devices less invasive and are best suited for realistic simulator applications, and can be used with alternative tracking technology such as electroencephalography (EEG) [42]. While the latter is less complex and can be mounted on a computer and it does not touch the user. Therefore, the visual space is already incorporated, the downside of this device is that it is fixed affecting data accuracy due to the unrestricted movements of the user's head.

Furthermore, there are two gaze estimation methods available to examine the geometry of the eye, the model-based method that use a priori knowledge to fit a 3D model to the eye image and the appearance-based method that attempts through a statistical model to train a mapping model of the eye gaze direction [43–48].

4.2 The need of a virtual pilot

In an air-to-air combat, the pilot's thought process is not linear and constrained by timed critical events. The continued accommodation of information requires the pilot to synthesize with sensemaking ability HMD imagery contributing toward correct decisions. In this perspective, the pilot has to be capable of acting adaptively when transitioning from one maneuver to another without missing important cues. This understanding involves multi-tasking and higher cognitive overload that could affect what makes up the pilot's mental model. Cognitive psychology defines that humans can perform up to two tasks at a time, while performing more tasks would result in reduced and a potential accident. In a combat scenario, the building blocks of the SA rely on sensors and communications. This all-encompassing information requires translating the holistic picture of detecting an adversary aircraft in a single fixation.

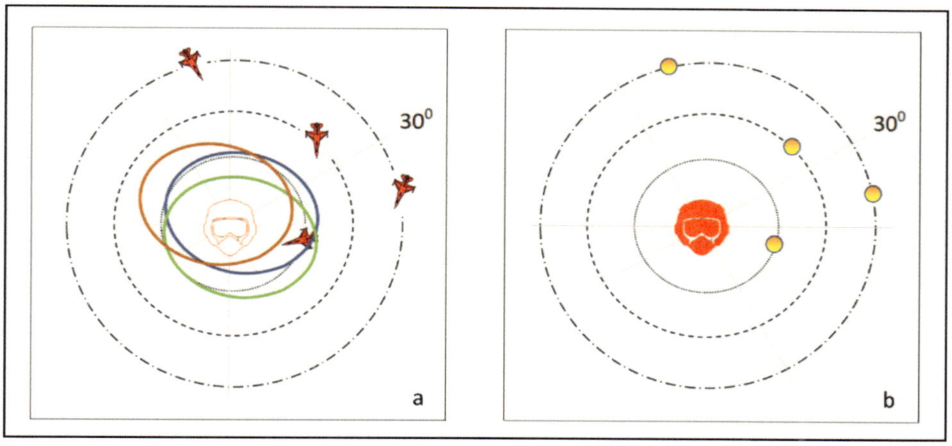

Figure 6.
The visual search limits: (a) cumulative probability of a pilot searching a 90° sector with 20 fixation/minute would detect an aircraft approaching direction by colored range (orange, green, blue). (b) The instant virtual map of the fixation points of the pilot's eye.

Fixation has been raised as a human factor red flag sometimes influencing modern-day accidents [49]. From the outset this cause is not new, there are recorded accidents from both military and civil aircraft directly tracing the cause of accidents to the fixation on the outside environment.

The latest 5th Generation of aircraft as the F-35 has departed from traditional mechanics of the F-15/F-16/F-21 multi-role aircraft with a new design philosophy. The pilot's HMD Gen III is essentially a component of the Information Fusion system, with the addition of advanced radar and infrared sensors that can pass data to a rocket battery on the ground to improve its accuracy. This capability allows the F-35 to track and destroy rockets, but to actually shut down a just launched one, the pilot would have to detect it just before reaching 100,000 ft [50]. In this case,

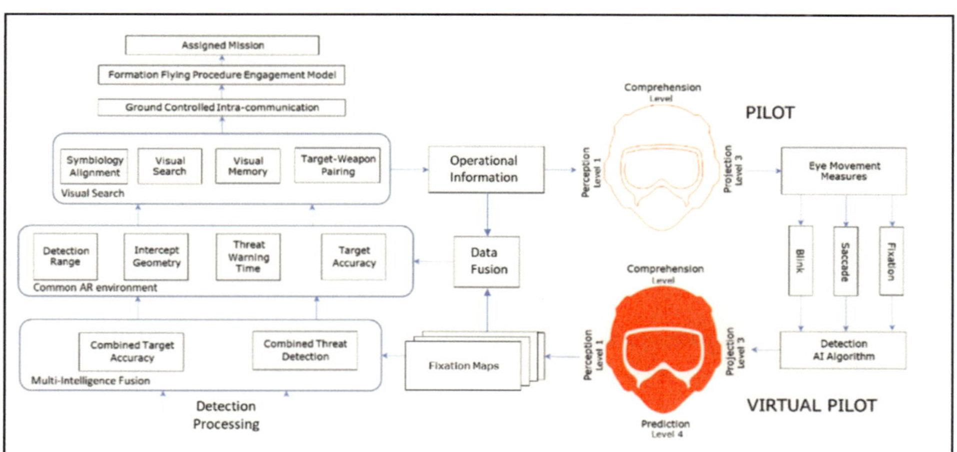

Figure 7.
Augment-me learning framework.

the pilot would have limited time to detect and intersect the rocket. This limitation highlights the pilot's uncued visual search limit. This limit is also evident against enemy aircraft or standard squadron formation. The systematic search in **Figure 6** highpoint this observation.

Figure 7 demonstrates that the pilot would have very limited time to respond to enemy attacks. This is the consequence of adversary aircraft falling just outside the pilot's central vision and the fixation at any given point. The established collaborative setting would require the use of AI to reproduce the pilot's fixations mechanisms generated in real-time during the mission.

5. The augment-me framework

The learning framework provides a cohesive structure for displaying information to pilots from on-board and off-board sensory sources while harvesting cognitive competencies of the pilot's decision-making when managing unexpected tactical maneuvers. The adaptation of this approach enhances precision engagements against fixed and moving targets as shown in **Figure 7**.

The learning framework was designed with the following objective:

- improve pilot's cueing ability
- add a prediction layer to the SA from acquiring long-range sensor data
- optimize the pilot visual search in combat by monitoring the pilot eye movement and pupil measures to design fixation maps
- enable AI technology to develop from fixation maps to tactical search areas
- combine AR and AI data to generate frozen search area sectors
- rapidly integration and transfer to ground support

The components required to build the framework are described as:

Assigned mission, engagement, intra-communication: High-level measures to quantify the combat capability to complete the mission.

Visual search: On-board, off-board measures visual performance and target prioritization.

Common AR engagement: Modular, distributed AR merging AI and virtual fixation environment (VFE).

Multi-intelligence fusion: Measure of intercept and detection probability quantifiable in a single variable.

Operational information: Imagery displayed on HMD and cockpit display.

Pilot HMD: Equipped with eye-tracking to find fixation points saccadic eye-motions.

Eye movement measures: Measures are identified using either are-based, dispersion-based or velocity-based algorithms.

Detection AI algorithm: Estimate the spatial position of the fixations.

Virtual pilot: Virtual agent or AI-powered processing for reconfigurable fixation maps.

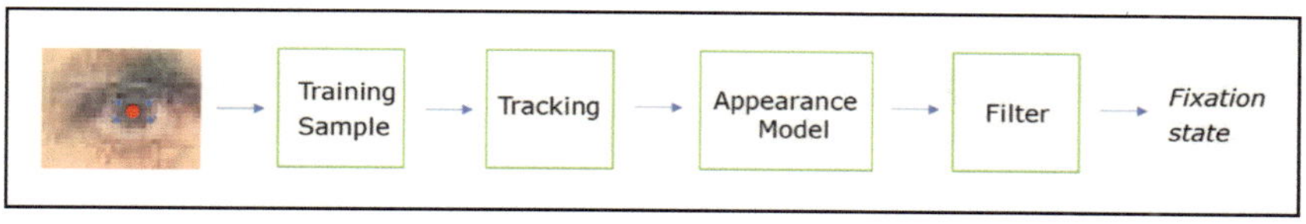

Figure 8.
Schematic of eye tracking filter structure.

Fixation maps: Fixation patterns for classification and maneuver profiles for ground-based tactical counteroffensives.

Since SA development requires recursive levels of abstractions and continued refinements of a combat scenario, the virtual pilot adds predictive action measures through the provision of virtual scan patterns, since each pattern allows a complete coverage of sliced sectors in **Figure 7b**, hence assuring a timely ground-based response. Efforts have been made to enhance the F-35 training pilots with eye tracking and EEG technology and to establish future applications [51].

5.1 AI augmented reality

Significant SA advantages can be gained through the implementation of AI-based solutions, especially in Level 3 SA. In this context, AI-augmented solutions provide assisted decision-making with real-time analysis and target prediction capability. The basic elements for online tracking combining optical flow, appearance model and a filter to determine the fixation state change are shown in **Figure 8**.

A technique to track eye motions is to use infrared light on the cornea and the pupil. A camera then is positioned to capture eye images and a processing algorithm estimates the center of the pupil and the positions of the glints. Pupil-cornea reflection measurements are performed in real-time, while the image processing algorithm provides an accurate measurement of the gaze. The appearance model updated only would function as a classifier while the filter is used to predict the search region and the fixation state can be modeled from Markov decision models, Cumulative Sum algorithms, Deep Learning and Machine Learning (ML) processes and Edge computing [52–54]. Edge computing can promote multimodal interaction with improved pilot-cockpit integration beyond standard interfaces with a high level of speed and accuracy [55]. The trend of applicable AI solutions could include Fog/Edge Computing to enable on-board computation rather than other off-board platforms, therefore improving SA responsiveness [24]. Further benefits of AR include low latency for data transmission from node to destination and no single-point failure. This follows the distributed approach of the Edge architecture that in case of a source failure, redirects instantaneously data transmission to an alternative edge network [56].

6. Conclusion

Modern combat aircraft are relying even more on robust sensor, weapon and network communication technology to provide pilots with superior SA. While this situation is evolving the pilot is expected to perform organized visual search patterns and avoid detection and tracking by enemy sensors. The visual search area and associated observation processes of the pilot have been utilized to model the aerial sensor footprint for threat detection. This data combined through an AR and AI learning framework is set to equip the pilot with improved visual search limits and tactical advantage.

Conflict of interest

The authors declare no conflict of interest.

Author details

Angelo Compierchio[1*], Phillip Tretten[1] and Prasanna Illankoon[2]

1 Luleå University of Technology, Luleå, Sweden

2 University of Moratuwa, Moratuwa, Sri Lanka

*Address all correspondence to: angelo.compierchio@ltu.se

References

[1] F-35A Lightning II [Internet]. 2023. Available from: https://www.af.mil/About-Us/Fact-Sheets/Display/Article/478441/f-35a-lightning-ii/ [Accessed: April 01, 2023]

[2] Unrivaled Capabilities [Internet]. 2023. Available from: https://www.f35.com/f35/about/5th-gen-capabilities.html [Accessed: April 04, 2023]

[3] Martinez-Conde S, Macknik SL, Hubel DH. The role of fixational eye movements in visual perception. Nature Reviews Neuroscience. 2004;**5**(3):229-240

[4] Stillion J. Trends in air-to-air combat: Implications for future air superiority. Center for Strategic and Budgetary Assessments [Internet]. 2015. Available from: https://csbaonline.org/uploads/documents/Air-to-Air-Report-.pdf [Accessed: February 16, 2023]

[5] Catching Up With F-35, Russia's Su-57 Fighters to Get 'Smarter' With AI-Enabled Sensor Fusion, Data Processing [Internet]. 2023. Available from: https://eurasiantimes.com/catching-up-with-f-35-russias-su-57-fighters-to-get-smarter/ [Accessed: April 06, 2023]

[6] Niehorster DC, Zemblys R, Beelders T, et al. Characterizing gaze position signals and synthesizing noise during fixations in eye-tracking data. Behavior Research Methods. 2020;**52**:2515-2534. DOI: 10.3758/s13428-020-01400-9

[7] Salvucci DD, Goldberg JH. Identifying fixations and saccades in eye-tracking protocols. In: Proceedings of the 2000 Association for Computing Machinery (ACM) Symposium on Eye Tracking Research & Applications. Florida, USA: Palm Beach Gardens; 2000. pp. 71-78

[8] Dehais F, Peysakhovich V, Scannella S, Fongue J, Gateau T. "Automation surprise" in aviation: Real-time solutions. In: Proceedings of the 33rd Annual Conference on Human Factors in Computing Systems. New York, USA: ACM; 2015. pp. 2525-2534

[9] Mannaru P, Balasingam B, Pattipati K, Sibley C, Coyne J. Cognitive context detection using pupillary measurements. In: Next-Generation Analyst IV. Vol. 9851. SPIE Defense and Security, Baltimore, MD, USA. 2016. pp. 244-251

[10] Velichkovsky BB, Khromov N, Korotin A, Burnaev E, Somov A. Visual fixations duration as an indicator of skill level in esports. In: Human-Computer Interaction–INTERACT 2019: 17th IFIP TC 13 International Conference, Paphos, Cyprus, September 2-6, 2019, Proceedings, Part I. Vol. 17. Lecture Notes in Computer Science. Vol. 11746. Cham: Springer; 2019. pp. 397-405

[11] Beringer D, Hancock PA. Exploring situational awareness: A review and the effects of stress on rectilinear normalization. In: Proceedings of the Fifth International Symposium on Aviation Psychology. Ohio State University, Department of Aviation Publishing. 1989. pp. 646-651

[12] Compierchio A, Tretten P. Human factors evaluation of shared real and virtual environments. In: Human Interaction, Emerging Technologies and Future Systems V: Proceedings of the 5th International Virtual Conference on Human Interaction and Emerging Technologies, IHIET 2021, August 27-29, 2021 and the 6th IHIET: Future systems (IHIET-FS 2021), October 28-30, 2021, Paris, France: Springer International Publishing; 2022. pp. 745-751

[13] Cutting JE. Images, imagination, and movement: Pictorial representations and their development in the work of James Gibson. Perception. 2000;**29**(6):635-648

[14] Dictionary of Military and Associated Terms [Internet]. 2019. Available from: https://www.jcs.mil/Portals/36/Documents/Doctrine/pubs/dictionary.pdf [Accessed: April 16, 2023]

[15] Li H, Zhang X, Shi G, Qu H, Wu Y, Zhang J. Review and analysis of avionic helmet-mounted displays. Optical Engineering. 2013;**52**(11):110901

[16] Previc FH, Ercoline WR, editors. Spatial Disorientation in Aviation. Reston, Virginia, USA: AIAA; 2004

[17] Melzer JE, Moffitt K. Head Mounted Displays. McGraw-Hill Publishing, the University of Michigan. 1997

[18] F-35: Under the Helmet of the World's Most Advanced Fighter. 2018. Available from: https://www.aviationtoday.com/2018/08/24/f-35-helmet-worlds-advanced-fighter/ [Accessed: April 08, 2023]

[19] F-35: Operational Based Vision Assessment (OBVA) "Human Vision Issues, Research and Future Research of the F-35 HMD. 2022. Available from: https://www.sto.nato.int›STO-EN-HFM-350 [Accessed: May 11, 2023]

[20] Joint Publication 3-09.3, Close Air Support, 25 November 2014 [Internet]. 2014. Available from: https://jdeis.js.mil/jdeis/new_pubs/jp3_09_3.pdf [Accessed: April 07, 2023]

[21] van Dijk L, Kiverstein J. Direct perception in context: Radical empiricist reflections on the medium. Synthese. 2021;**198**:8389-8411. DOI: 10.1007/s11229-020-02578-3

[22] Gibson JJ. The Senses Considered as Perceptual Systems. Boston, USA: Houghton Mifflin Company; 1966

[23] Wickens CD. Pilot attention and perception and spatial cognition. In: Human Factors in Aviation and Aerospace. London, UK: Academic Press; 2023. pp. 141-170. DOI: 10.1016/B978-0-12-420139-2.00009-5

[24] Munir A, Aved A, Blasch E. Situational awareness: Techniques, challenges, and prospects. AI. 2022;**3**(1):55-77

[25] Carmigniani J, Furht B, Anisetti M, Ceravolo P, Damiani E, Ivkovic M. Augmented reality technologies, systems and applications. Multimedia Tools and Applications. 2011;**51**:341-377

[26] Endsley MR. Toward a theory of situation awareness in dynamic systems. Human Factors. 1995;**37**(1):32-64

[27] Bayle E, Guilbaud E, Hourlier S, Lelandais S, Leroy L, Plantier J, et al. Binocular rivalry in monocular augmented reality devices: A review. Situation Awareness in Degraded Environments. 2019;**2019**(11019):136-149

[28] Yildirim I, Schneider KA. Neural dynamics during binocular rivalry: Indications from human lateral geniculate nucleus. Eneuro. 1 Jan 2023;**10**(1). DOI: 10.1523/ENEURO.0470-22.2022

[29] Dempo A, Kimura T, Shinohara K. Perceptual and cognitive processes in augmented reality–comparison between binocular and monocular presentations. Attention, Perception, & Psychophysics. 2022;**84**(2):490-508. DOI: 10.3758/s13414-021-02380-4

[30] Cao T, Wang L, Sun Z, Engel SA, He S. The independent and shared mechanisms of intrinsic brain dynamics: Insights from bistable perception. Frontiers in Psychology. 2018;**9**:589

[31] Wilson AD, Golonka S. Embodied cognition is not what you think it is. Frontiers in Psychology. 2013;**4**:58

[32] Schallhorn S, Daill K, Cushman WB, Unterreiner R, Morris A. Visual Search in Air Combat. Pensacola, FL: Naval Aerospace Medical Research Lab, NAMRL Publications; 1990

[33] Walter K, Bex P. Cognitive load influences oculomotor behavior in natural scenes. Scientific Reports. 2021;**11**(1):12405

[34] Klein G, Drummond T. Robust visual tracking for non-instrumental augmented reality. In: The Second IEEE and ACM International Symposium on Mixed and Augmented Reality, 2003. Proceedings. Tokio, Japan: IEEE; 2003. pp. 113-122

[35] Corbett M, Shang J, Ji B. GazePair: Efficient pairing of augmented reality devices using gaze tracking. IEEE Transactions on Mobile Computing. 2023

[36] Stone A, Rajeev S, Rao SP, Panetta K, Agaian S, Gardony A, et al. Gaze depth estimation for eye-tracking systems. In: Multimodal Image Exploitation and Learning 2023. Vol. 12526. Orlando, Florida, USA: SPIE; 2023. pp. 143-152

[37] Shree DVJ, Murthy LR, Saluja KS, Biswas P. Operating different displays in military fast jets using eye gaze tracker. Journal of Aviation Technology and Engineering. 2018;**8**(1):31

[38] Lutnyk L, Rudi D, Schinazi VR, Kiefer P, Raubal M. The effect of flight phase on electrodermal activity and gaze behavior: A simulator study. Applied Ergonomics. 2023;**109**:103989

[39] Reis GA, Miller ME, Geiselman EE, Langhals BT, Kabban CM, Jackson JA. Effect of visual field asymmetries on performance while utilizing aircraft attitude symbology. Displays. 2023;77:102404

[40] Li W-C, Lin JJ, Braithwaite G, Greaves M. The development of eye tracking in aviation (ETA) technique to investigate pilot's cognitive processes of attention and decision-making. In: Proceedings of the 32nd Conference of the European Association for Aviation Psychology (EAAP) Publishing,, Cascais, Portugal, 26-30 September 2016

[41] Dehais F, Behrend J, Peysakhovich V, Causse M, Wickens CD. Pilot flying and pilot monitoring's aircraft state awareness during go-around execution in aviation: A behavioral and eye tracking study. The International Journal of Aerospace Psychology. 2017;**27**(1-2):15-28

[42] Different Kinds of Eye Tracking Devices. 2020. Available from: https://www.bitbrain.com/blog/eye-tracking-devices [Accessed May 21, 2023

[43] Babu MD, JeevithaShree DV, Prabhakar G, Saluja KPS, Pashilkar A, Biswas P. Estimating pilots' cognitive load from ocular parameters through simulation and in-flight studies. Journal of eye movement. Research. 2 Sep 2019;**12**(3):10. DOI: 10.16910/jemr.12.3.3

[44] Klaproth OW, Halbrügge M, Krol LR, Vernaleken C, Zander TO, Russwinkel N. A neuroadaptive cognitive model for dealing with uncertainty in tracing pilots' cognitive state. Topics in Cognitive Science. 2020;**12**(3):1012-1029

[45] Gomolka Z, Kordos D, Zeslawska E. The application of flexible areas of interest to pilot mobile eye tracking. Sensors. 2020;**20**(4):986

[46] Naeeri S, Mandal S, Kang Z. Analyzing pilots' fatigue for prolonged flight missions: Multimodal analysis approach using vigilance test and eye tracking. In: Proceedings of the Human Factors and Ergonomics Society Annual Meeting. Vol. 63(1). Los Angeles, CA: SAGE Publications; 2019. pp. 111-115

[47] An Eye Tracking based aircraft helmet mounted display aiming system. 2022. Available from: https://www.techrxiv.org/articles/preprint/An_Eye_Tracking_based_Aircraft_Helmet_Mounted_Display_Aiming_System/18093233 [Accessed: June12, 2023]

[48] Modi N, Singh J. A review of various state of art eye gaze estimation techniques. Advances in Computational Intelligence and Communication Technology: Proceedings of CICT. 2019;**2021**:501-510

[49] Pilot Duty of Care and the Role of the Human Factors Expert. 2014. Available from: https://www.meaforensic.com/pilot-duty-of-care-and-the-role-of-the-human-factors-expert/ [Accessed: April 06, 2023]

[50] Could the F-35 Really Shoot Down an Enemy Ballistic Missile? 2020. Available from: https://nationalinterest.org/blog/buzz/could-f-35-really-shoot-down-enemy-ballistic-missile-135442 [Accessed: April 20, 2023]

[51] Carroll M. Enhancing HMD-based F-35 training through integration of eye tracking and electroencephalography technology. In: Schmorrow DD, Fidopiastis CM, editors. Foundations of Augmented Cognition. AC 2013, Lecture Notes in Computer Science. Vol. 8027. Berlin, Heidelberg: Springer; 2013. DOI: 10.1007/978-3-642-39454-6_3

[52] Klaib AF, Alsrehin NO, Melhem WY, Bashtawi HO, Magableh AA. Eye tracking algorithms, techniques, tools, and applications with an emphasis on machine learning and internet of things technologies. Expert Systems with Applications. 2021;**166**:114037

[53] Yang T, Cappelle C, Ruichek Y, El Bagdouri M. Online multi-object tracking combining optical flow and compressive tracking in Markov decision process. Journal of Visual Communication and Image Representation. 2019;**58**:178-186

[54] Gunawardena N, Ginige JA, Javadi B. Eye-tracking technologies in mobile devices using edge computing: A systematic review. ACM Computing Surveys. 2022;**55**(8):1-33

[55] Edge Computing Poised to Give AR Wearables a Big Boost. 2021. Available from: https://www.fierceelectronics.com/electronics/edge-computing-poised-to-give-ar-wearables-a-big-boost [Accessed: May 02, 2023]

[56] Letaief KB, Shi Y, Lu J, Lu J. Edge artificial intelligence for 6G: Vision, enabling technologies, and applications. IEEE Journal on Selected Areas in Communications. 2021;**40**(1):5-36

Chapter 2

A Wearable Force-Feedback Mechanism for Immersive Free-Range Haptic Experience

Peter Kudry and Michael Cohen

Abstract

This chapter presents the development of a wearable force-feedback mechanism designed to provide a free-range haptic experience within the spectrum of Extended Reality (XR). The proposed system offers untethered six degrees-of-freedom and small- to medium-scale force-feedback, enabling users to immerse themselves in haptic interactions within virtual environments. The hardware comprises a modified 3D Systems Touch haptic device, driven by software that allows for ambulatory exploration of various haptic aspects. Two experiments were conducted to evaluate the precision, ergonomics, stability, usability, user experience, and performance of the system. Despite indication of software and hardware deficiencies, the results highlight the potential of combining haptic force-feedback and ambulatory XR to enhance immersion in free-range virtual environments. Furthermore, the integration of Mixed Reality pass-through enables users to seamlessly merge real-world environments with augmenting virtual elements. This extension contributes to the exploration of new possibilities for immersive and interactive experiences within mixed reality applications. Future research can delve deeper into the prototype's potential, further unlocking opportunities for haptic-enabled ambulatory XR experiences and pushing the boundaries of immersive technologies.

Keywords: haptic interface, wearable computer, virtual reality, augmented reality, mixed reality, human-computer interaction (HCI), ambulatory application, force-feedback, perceptual overlay, tangible user interface (TUI), spatial computing, multimodal interaction, extended reality (XR), mixed reality (MR), virtual environments (VE)

1. Introduction

In this section, we survey the evolution of extended reality (XR), discuss its present trends, highlight areas that are comparatively less developed within this paradigm, emphasize the significance of further advancements, and provide an overview of the alternate solution we have developed through our research.

1.1 History of XR

Origins of contemporary virtual reality (VR) can be traced back to the early twentieth century with "Link Trainer," the first flight simulator, developed in 1929 and patented in 1931 [1]. However, VR hardware primarily targeted commercial use, and early attempts to bring this technology to the general public for entertainment purposes encountered commercial failures. Disappointments include the Sega VR-1 and Nintendo Virtual Boy, released in 1994 and 1995, respectively. This situation has significantly changed in the recent years, as marked by the release of the Oculus Rift in 2010, which sparked a renaissance in VR.

Similarly, in the augmented reality (AR) domain, there were already several established AR solutions by 2010, but the ARToolKit, developed in 2000 [2], can be considered the precursor to modern AR. By 2020, use of VR and AR, collectively referred to as extended reality (XR), expanded across various domains, including entertainment, education, and industrial training. XR technologies have become widely accessible to the general public through integration with popular gaming consoles, smartphones, and stand-alone headsets, offering immersive experiences at a fraction of the cost compared to earlier VR waves [3, 4]. Such rapid technological advancements have exponentiated numerous trends within a relatively short period.

1.2 Ongoing and emerging trends in XR

In recent years, integration of VR and XR across various industries, including marketing, design, and retail, has caused a notable shift in dominance within specific demographic segments [5]. This shift is evident in the current landscape of XR experiences, which predominantly cater to industrial applications rather than gaming and entertainment. The 2020 XR Industry Insight report reveals that about ⅔ of AR development companies are focused on industrial applications, while consumer products account for only about ⅓ [6]. This trend can be attributed to the unique ability of XR technologies to virtualize environments and scenarios that are typically expensive or hazardous.

The healthcare industry has embraced XR technologies as utilized in therapeutic interventions for phobias and anxiety disorders, as well as in assisting individuals with autism in developing social and communication skills. AR, in particular, holds immense potential as a valuable visual aid for surgical procedures.

These ongoing and future trends in XR are propelled by advancements in hardware. XR devices are continuously shrinking in size, becoming more portable, and gaining enhanced processing power, thereby improving accessibility and ergonomics. The shift toward self-contained and untethered devices, exemplified by products like Meta Quest, contribute to an enhanced user experience [7]. Another significant advancement is the widespread deployment of 5G mobile networks, offering the potential for improved VR streaming and seamless collaborations within virtual environment (VE) by delivering higher bandwidth and lower latency.

In summary, the application of VR and XR has resulted in a transformative shift. Industrial applications currently dominate XR experiences, with healthcare serving as a prominent example of XR technology adoption. Ongoing advancements in hardware, including smaller, more powerful, and self-contained devices like the Apple Vision Pro, and the promise of mobile broadband connectivity, contribute to the continuous growth and development of XR applications.

1.3 Importance of haptic feedback in immersive environments

Significant advancements have occurred in haptics for VR since the release of the first Oculus Rift development kit a decade ago. One notable improvement is the transition from localized, three degrees-of-freedom models to room-scale, six degrees-of-freedom tracking. This advancement allows complete tracking of user movement, including wireless tracking of hand-held controllers. Modern controllers now include capacitive touch sensors that detect finger movements, enhancing user interaction with virtual environments. The tactile representation of buttons on controllers reinforces immersion by providing coherent sensory experience. However, in terms of haptic feedback from virtual objects, progress has been limited, as most controllers still rely on vibration feedback. Although solutions such as gloves and full-body suits provide more realistic haptic feedback, they are not widely accessible due to their cost and specialized markets [8, 9].

Sense of presence, the subjective perception of truly being in a virtual environment, is a crucial aspect of immersive VR experiences [10, 11]. While realistic visual environments can can foster suspension of disbelief, lack of coherent haptic feedback when interacting with the VE disrupts the illusion. Passive haptics, which involve using physical props placed in the real space, can partially address this issue. For example, placing a box in the real space allows a user to lean on it in the game, creating the haptic illusion of taking cover. However, this approach is impractical as it requires adapting the real space to match the virtual environment (VE). Another solution, known as active haptics, utilizes robotics to dynamically position props based on user actions. Although this approach overcomes the flexibility issue, it may introduce latency, which can be problematic in fast-paced gaming scenarios [12, 13].

Despite the challenges, several studies have shown that haptic feedback improves interaction, spatial guidance, learning, and sense of presence in VR environments [9, 14, 15]. However, current systems still struggle to provide high-quality force or tactile feedback [16]. Achieving relevant and believable combination of small-scale haptic feedback, such as texture simulation, and medium-scale force-feedback, poses challenges in terms of form-factor, ease of deployment, and precision [17]. These challenges are particularly evident in wearable solutions that aim to immerse users in a virtual environment through VR. In conclusion, haptic feedback, along with the stimulation of other senses beyond visual and auditory, significantly impacts quality of immersion in VR. The greater the variety of sensory feedback, the deeper the immersion [18, 19].

1.4 Recent haptic feedback solutions for immersive experiences

Despite challenges in implementing haptic feedback devices, researchers are actively working on solutions to overcome these obstacles and drive advancements in this field. This section provides a brief description of various devices that have recently addressed some of the aforementioned issues.

1.4.1 Cross-field haptics: push-pull haptics combined with magnetic and electrostatic fields (2015)

The concept of cross-field haptics involves the use of multifield physical quantities to replicate various textures. One approach is to utilize magnetorheological fluid (MRF) positioned between an array of electromagnets and conductive electrodes.

The behavior of MRF is influenced by the magnetic field generated by these layers. When no magnetic field effects are present, MRF behaves as a Newtonian fluid, but its viscosity changes in response to variations in a field, transforming it into a non-Newtonian fluid. In simpler terms, the viscosity of the MRF can be adjusted based on the strength of a magnetic field produced by coils, allowing for simulation of various textures [20].

1.4.2 Magic table: deformable props using Visuo haptic redirection (2017)

The Magic Table employs haptic retargeting using a single physical object to represent multiple objects in a VE [21]. Specifically, it utilizes body warping and the technique of redirected walking. Body warping refers to perceived change in the shape of objects, and redirected walking introduces additional translations and rotations to a head-mounted display (HMD), causing users to physically traverse paths that differ from their virtual perception.

1.4.3 HaptoBend: utilizing shape-change to enhance virtual reality (2017)

HaptoBend is a passive haptic feedback device that enables users to experience the transformation of a flat 2D object into a multi-sided 3D object through bending [22]. It uses four ridged sections with hinged connections, allowing users to deform it into preferred physical handform shapes and interact with virtual objects in Unity for use in VR.

1.4.4 AirPiano: enhancing music playing experience in virtual reality with mid-air haptic feedback (2017)

AirPiano is a unique haptic feedback device that replicates the experience of playing a piano by creating touchable keys in free space using ultrasonic vibrations [23]. While this specific simulation is not generalizable, it demonstrates the potential of using ultrasonic acoustic waves for haptic feedback in specialized applications.

1.4.5 CLAW: a multifunctional handheld VR haptic controller (2018)

CLAW is a virtual reality controller that enhances traditional controller capabilities by providing force-feedback and actuated movement specifically to the index finger [24]. It aims to provide feedback for three different types of interactions: touching, grasping, and triggering. Vibrations are used to simulate textures when touching, and a servomotor is employed for the grasping and triggering actions.

1.4.6 Haptic revolver: touch, shear, texture, and shape rendering on a VR controller (2018)

The Haptic Revolver shares similarities to the CLAW in terms of features but employs a different approach to achieve the same functionality. It utilizes an actuated wheel that moves up and down beneath the user's finger to create contact with a virtual surface [25]. As a user's finger glides along a surface, the wheel spins to generate shear forces and motion feedback. Unlike the CLAW, the Haptic Revolver offers the advantage of user-interchangeable wheels, which can provide various textures, shapes, edges, and active elements to enhance haptic experience.

1.4.7 Wearable fingertip haptic device for remote palpation: Characterization and interface with a virtual environment (2018)

The wearable fingertip haptic device consists of two primary subsystems that enable simulation of palpation (feeling a shape) of virtual object presence and surface stiffness [26]. The first subsystem includes an inertial measurement unit, which tracks one's finger's motion and adjusts the linear displacement of a pad toward the fingertip. The second subsystem controls the pressure in a variable compliance platform using a motorized syringe, allowing for simulation of surface stiffness when touched.

1.4.8 Interactive sculpting using augmented-reality, mesh morphing, and force-feedback: Force-feedback capabilities in an augmented reality environment (2018)

This interactive sculpting prototype integrates force-feedback capabilities with AR to facilitate realtime morphing of geometric surfaces. Its purpose is to provide designers with a direct way to modify component shapes through interaction with virtual representations [27]. The system utilizes a radial basis function (RBF) morpher for realtime computations. It incorporates a camera as an input device and an HMD with OLED screens as an output, effectively creating an AR helmet. The Geomagic Touch X device functions as a haptic interface, enabling users to manipulate virtual objects and experience force-feedback along three spatial directions using its motors.

1.4.9 Muscleblazer: force-feedback suit for immersive experience (2019)

The Muscleblazer suit is a lightweight exo-suit designed for force-feedback in VR [28]. It utilizes solenoid valves and micro-controller boards to activate Pneumatic Gel Muscles (PGMs) and generate flexible and lightweight forces. In conjunction with a VR game, users can engage in shooting enemies using an HTC VIVE controller, the PGMs providing haptic display upon being shot. The suit is adaptable to both VR and AR environments, enabling wireless communication for the generation of force-feedback effects.

1.4.10 Wireality: enabling complex tangible geometries in virtual reality with worn multi-string haptics (2020)

Wireality aims to overcome various challenges in haptic feedback technology. It is a self-contained wearable system that enables precise positioning of individual hand joints in three-dimensional space, using retractable wires that can be programmatically locked [29]. This allows for realistic interactions with intricate geometries, such as wrapping fingers around objects such as railings. The device is lightweight, comfortable, and durable, while also being affordable with production cost less than $50 USD.

1.4.11 Wrist-worn prototypes for XR input and haptics by meta (2021)

Facebook is actively exploring haptic technology through wrist-worn input devices. These devices are still in the early prototype stage, but Facebook envisions using electromyography (EMG) sensors to track button presses and enable keyboard-less typing by sensing electrical signals in a user's arm. Previous prototypes from Facebook have included wristbands with inflatable bladders to apply pressure on one's wrist and vibrating actuators for vibro-tactile feedback. Facebook's heavyweight involvement in haptics suggests significant advancements in the field may be on the horizon [30].

1.4.12 QuadStretch: a forearm-wearable multi-dimensional skin stretch display for immersive VR haptic feedback (2022)

QuadStretch is a newly developed forearm-worn device designed for VR interaction. It features a compact and lightweight design and utilizes counter-stretching to stretch a user's skin [31]. This is achieved by moving in opposite direction a pair of tactors, secured to one's skin surface with an elastic band. QuadStretch offers various VR interaction scenarios that showcase its unique characteristics, including intensity-based activities such as boxing and pistol shooting, passive tension and spatial multi-dimensionality in activities like archery and slingshot, and continuity in complex movements like flying and climbing.

1.4.13 Free-range haptic immersive XR (2022)

While most solutions in this section do not provide unrestricted mobility and positional displacement for haptic interfaces with force-feedback in virtual environments, our own project addresses this limitation. We have developed hardware by modifying the 3D Systems Touch haptic device, integrated with software using the Unity game engine and low-level device drivers [32]. This combination enables medium-scale force display in mobile XR applications, allowing enhanced interaction and mobility within virtual environments.

2. Materials and methods

2.1 Problem description

In Ref. [16], a paradigm shift in human-computer interaction was described, highlighting emergence of devices like the force-feedback haptic stylus. The 3D Systems Touch haptic devices are particularly relevant, offering precise tracking and possible support among major 3D design tools such as Blender, Maya, and 3DS Max. These stylus devices, designed for desktop use, are ideal for immersive CAD experiences due to their accurate force-feedback and high precision.

While there are other wearable force-feedback solutions available, such as the TactGlove [33], SenseGlove Nova [34], and Power Glove [35], stylus-based interfaces have the advantage of being grounded and providing force-feedback in space. However, they impose constraints on the user's arm movement as the stylus interacts with virtual surfaces. This limitation does not apply to other devices that offer the ability to sense size, shape, stiffness, and motion of virtual objects, but lack positional translation.

Wireality (§1.4.10), a device comparable in functionality to the Touch stylus, was released in 2020. While Wireality effectively addresses many challenges outlined regarding the importance of haptic feedback, its precision may not meet the requirements of 3D modeling and sculpting, and it does not support texture simulation or vibrotactile feedback.

2.2 Prototype design overview

Hence, we took on a mission of transforming a stylus device designed for desktop use into a wearable haptic interface capable of delivering untethered six degrees-of-freedom in a VR setting. This endeavor spawned various challenges, including addressing the ergonomics of a wearable harness, managing weight and power considerations, and overcoming limitations inherent to the haptic device itself [32, 36].

2.2.1 Hardware

Harness—Upon careful consideration and subsequent dismissal of over-the-shoulder arrangements, the decision was made to develop an adjustable platform where the stylus base could be mounted in front of the user. The key component is a user-worn vest, which must possess sufficient strength to bear the weight of all the hardware without compromising comfort or impeding natural movement. A tactical vest initially designed for survival games emerged as a versatile choice. To accommodate the necessary modifications, mounting clips for two sheets of 3-mm-thick aluminum were added to both the front and back sections of the vest, as shown in **Figure 1**.

Front and rear plating—Various alternatives for the plating material, such as acrylic or thinner steel, were considered. However, these materials presented drawbacks: Acrylic and similar options were found to be lightweight and flexible, while thicker steel plates were heavy and rigid. Neither of these extremes were suitable for the intended use case. In contrast, aluminum has properties that strike a balance between the two, so was deemed to be the appropriate choice.

The aluminum plate on the back serves the purpose of mounting a full-size 15.6″ laptop, which powers the Meta Quest 2 HMD in "Quest Link" mode. This configuration ensures optimal performance, as the Quest 2 is based on the Android mobile OS platform with hardware that imposes performance limitations. Since Open Haptics,

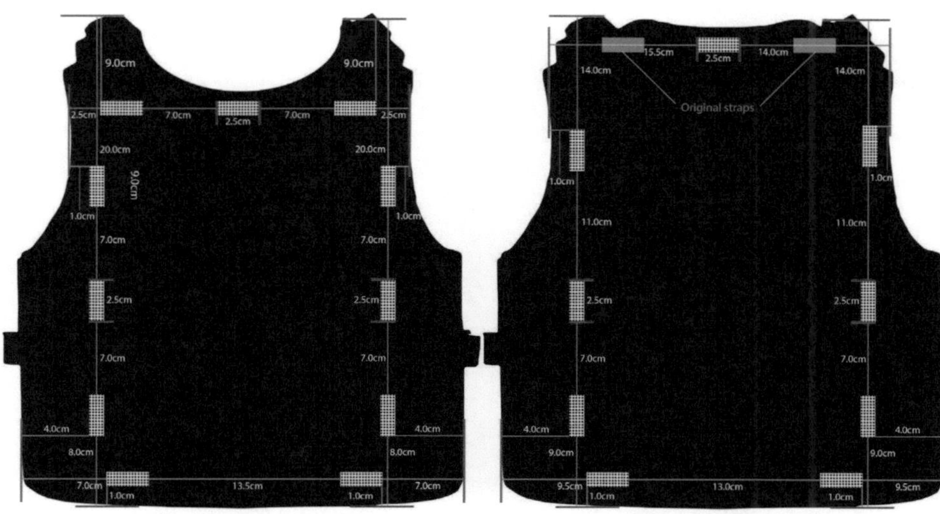

A: Strap positions, front **B: Strap positions, back**

Figure 1.
Tactical vest customized for supporting force-feedback device platform: Strap positions, front (A) and back (B).

the software development kit (SDK) for 3D Systems devices, is not compatible with Android, it is necessary to drive the haptic device separately.

3D Systems Touch platform—A perpendicular attachment is made on the front aluminum plate, utilizing another 3-mm-thick plate as the base for the stylus assembly, which acts as a cantilever for the wearable device. In the middle of this front aluminum plate, a channel is cut along its length, allowing the stylus to be adjusted closer to or further away from the user's torso. This ergonomic feature accommodates users of varying height and arm length, ensuring a comfortable fit. To ensure electrical insulation between the aluminum and electronic components of the assembly, the stylus device itself is mounted on a sheet of laser-cut 3 mm-thick acrylic.

Stylus assembly modifications—Modifications were also necessary for the stylus assembly specifically tailored to this application. Originally, the base of the Touch

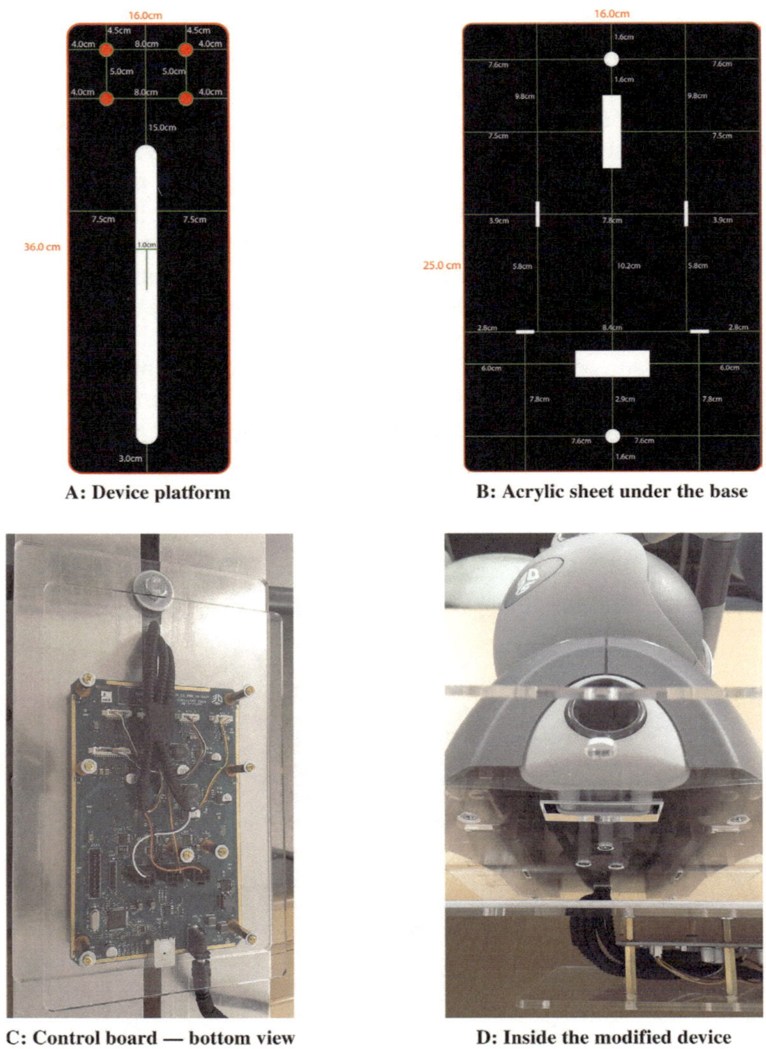

Figure 2.
Force-feedback device modifications: (A) device platform; (B) acrylic sheet under the base; (C) control board—bottom view; and (D) inside the modified device.

stylus contained weights to prevent it from tipping during desktop use. These weights were removed, as the cantilever provides secure enough mounting for the user to perceive no vertical flexing of the shelf and almost no horizontal flex. Further adjustment included moving the Touch main board from the bottom of the stylus's base assembly to the bottom of the shelf while preserving the adjustment feature. These (warranty-voiding) modifications and platform structure are shown in **Figure 2**.

Power delivery—The wearable assembly comprises three primary devices, necessitating the development of suitable power delivery systems. The power supplies for the HMD and laptop were utilized in original arrangement, as they come equipped with their own internal batteries. However, power for the stylus servomotor is supplied by an external power bank securely attached to the user's waist. **Figure 3** shows an abstract representation of the entire system, while **Figure 4** shows actual use.

2.2.2 Software

The implemented software is divided into two main subsystems, as shown in **Figure 5**.

Virtual reality—The immersive environment is streamed from the laptop, mounted on the rear of the vest, to the HMD. As mentioned earlier, the Meta Quest 2 is a standalone device based on the Android platform. However, due to incompatibility with the Open Haptics SDK, the Quest must be operated in Link mode, which transforms it into a tethered HMD. Unlike some other HMDs, the Quest 2 utilizes inside-out tracking, eliminating need for stationary "lighthouse" sensors in the user's space, as required by older devices such as the HTC Vive or Oculus Rift.

The VR environment is created in the Unity game engine, leveraging the Oculus XR plugin and the Mixed Reality Toolkit (MRTK). The Oculus XR plugin handles

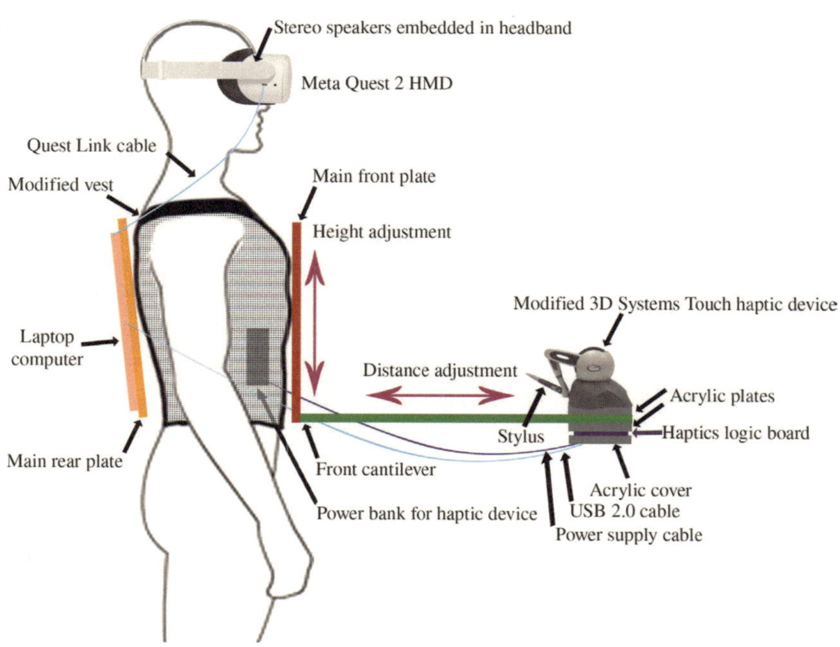

Figure 3.
Hardware assembly as worn—Side view (profile).

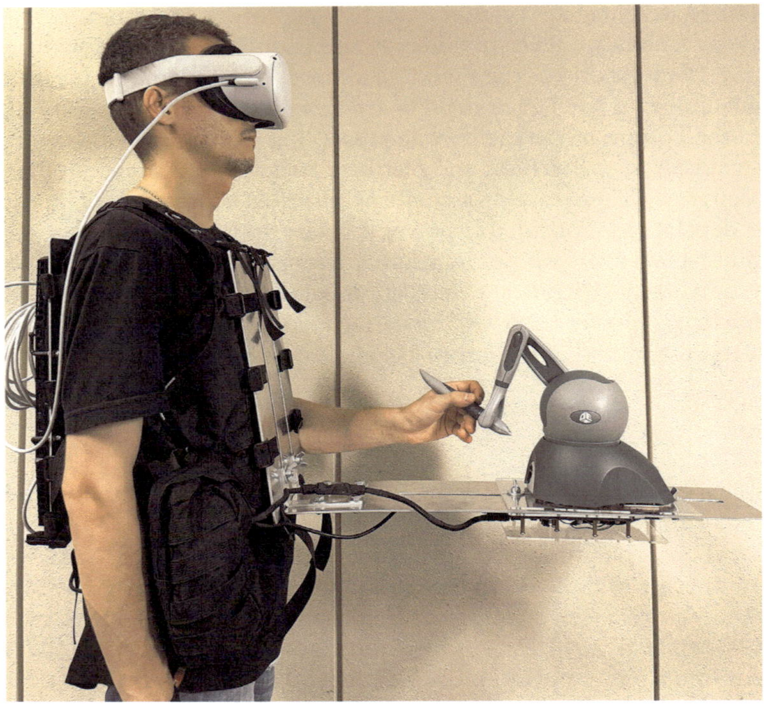

Figure 4.
Hardware assembly as worn—Photo.

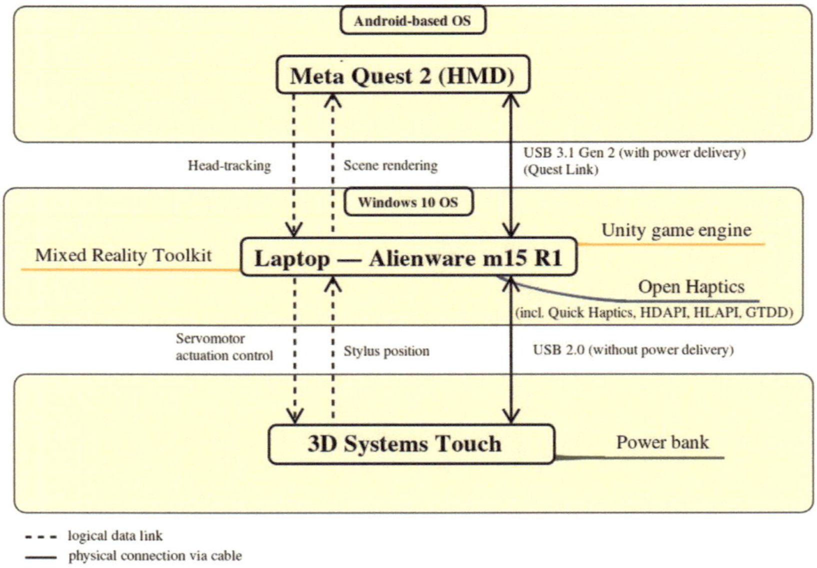

Figure 5.
Wireline software architecture.

lower-level functionality such as stereoscopic rendering for the HMD, the Quest Link feature (which essentially turns the Quest into a thin client of the PC), and input subsystems that provide controller support and HMD tracking [37]. The MRTK encapsulates these low-level features and extends them with hand-tracking capabilities and other features such as gesture-specified teleportation, ray-cast reticle operation, and physics-enabled hand models [38].

Haptic force-feedback—Software control of the 3D Systems stylus is facilitated through the Open Haptics for Unity plugin, which allows for integration of various 3D haptic interactions in the Unity environment. The plugin comprises several components, including the Quick Haptics micro API, Haptic Device API (HDAPI), Haptic Library API (HLAPI), Geomagic Touch Device Drivers (GTDD), and additional utilities [39, 40].

The structure of the Open Haptics plugin for Unity differs from the native version. Instead of using Quick Haptics for haptics and graphics, this package employs the `OHToUnityBridge` DLL to establish communication between the Unity controller, the HapticPlugin script written in C#, and the HD/HL APIs written in the C programming language. Analysis of the dependencies of `OHToUnityBridge.dll` revealed that this library directly invokes the HD, HL, and OpenGL libraries, without relying upon Quick Haptics [39, 40].

2.2.3 Environment description

The rest of this section outlines the four primary components of the engineered prototype [32, 36].

1. Scene loader—The Unity scene labeled as the "scene loader" is not directly accessible for selection by the user. Instead, it functions as a container, encompassing not only objects from the sub-scenes but also serving as a wrapper for the application's scene management system.

2. Scene selector—As illustrated in **Figure 6**, upon launching the demo application, the initial "splash" scene known as the "Scene Selector" is loaded additively into the Scene Loader. This scene includes various selectors that gather information regarding intended use of the main sub-experiences. The user is presented with pillars topped with buttons, a lever, and a canvas containing touchable User Interface (UI) buttons.

   The lever serves the purpose of indicating the user's chirality, determining the dominant hand to be used with the haptic stylus. On the canvas, there are interactible entries representing previously saved states of sculpting and carving sessions. When a session concludes, its state is saved and can later be reloaded, allowing the user to resume exploration.

   The left button, labeled "Haptic Sandbox," is used to load one of the sub-experiences, while the button labeled "Sculpting & Carving" is used to load the other sub-experience, which purpose is further explained below. In this scene, the user relies on hand tracking, gestures, and physics to interact with the virtual environment. Additionally, a palm-up "put-myself-there" gesture teleports the user within the designated play area.

3. Haptics sandbox—As shown in **Figure 7**, the "haptics sandbox" offers the user a range of haptic simulations comprising five distinct stages, showcasing different capabilities of the haptic mechanism.

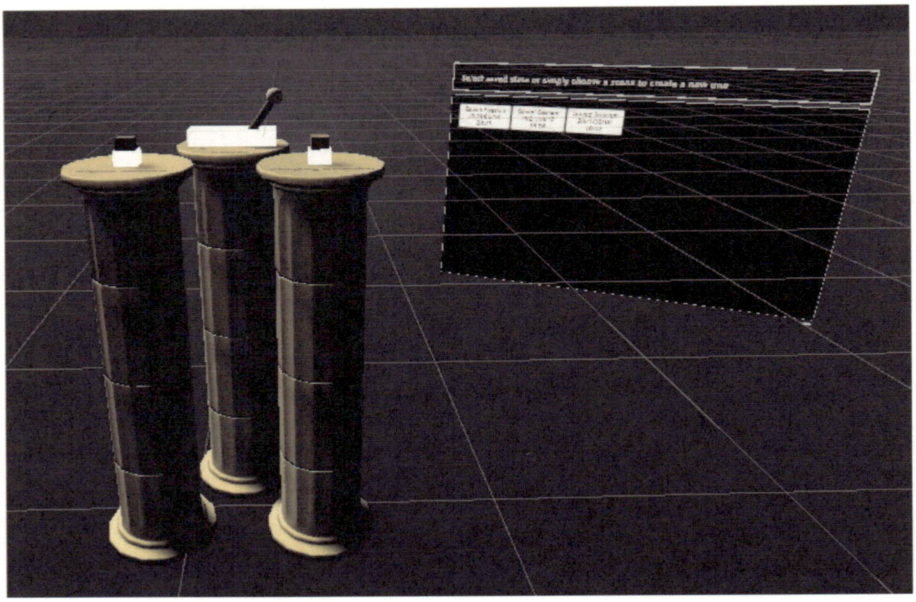

Figure 6.
Scene selector (splash scene).

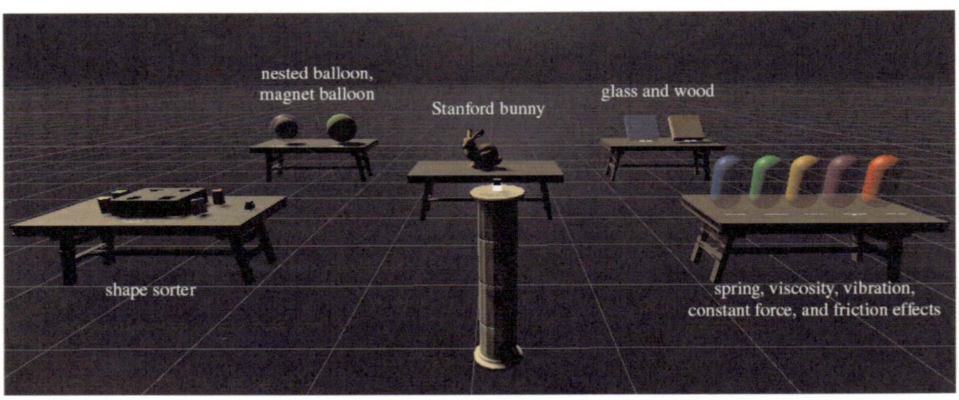

Figure 7.
Haptics sandbox scene.

The first stage presents a block shape-sorting experience. Users can teleport to an area in front of a table with various shapes to be inserted into corresponding cutouts in a pre-cut recepticle. Each block has a unique shape and a unique correct cutout, resembling a children's toy. Positioned comfortably in front of the desk, users can grasp the physical haptic stylus, mirrored by a congruent shape in the virtual space. Real physical movements are then projected into the virtual environment. When users palpate a "touchable" object in the virtual space, the haptic device responds by mechanically locking or constraining movement and rotation, simulating contact with a physical object. Additionally, blocks can be picked up by pressing a button on the

DOI: http://dx.doi.org/10.5772/intechopen.1002679

stylus, virtually grabbing them, and lifting them. Simulation of weight and mass of virtual objects is achieved by the haptic plugin's ability to translate Unity material physical properties into attributes and parameters processed by the haptics engine, which controls the servomotors in the assembly.

Another stage features two balloon-like sculptures. One sculpture consists of outer and inner spheres. The outer layer represents a relatively weak material that can be punctured with a certain amount of force, akin to popping a balloon with a pin. Once the outer layer is pierced, the stylus encounters a solid, impenetrable inner sphere, allowing users to feel its shape across the surface. When users want to pull back out of the outer layer, they must exert the same amount of force as when popping in. The other sculpture, instead of resisting touch, attracts the stylus to its surface and restricts stylus movement to match its shape. This sensation is akin to dragging a magnetic stick over a metallic surface. Sliding the stylus tip across the surface is effortless, but to detach it, a certain amount of force must be applied to overcome the "stickiness." If the force exceeds the virtual-magnetic attraction, the stylus breaks away from the spherical shape.

At the next stage table, users are presented with two angled boards, representing virtual materials simulating glass and wood. This experience highlights the haptic's ability to simulate textures and smoothness. By teleporting to the desk area, users can use the stylus to touch these two boards, comparing the tactile sensations.

Lastly, a table featuring five differently colored capsules is presented to the user. Each capsule represents a distinct tactile effect, allowing users to experience elastic springiness, viscosity, vibration, constant force, and friction effects. As a bonus feature, users also have the opportunity to palpate the Stanford bunny, a commonly used model for benchmarking 3D software due to its representative nature.

4. Sculpting and carving—The second available space for the user to enter is the "Sculpting and Carving" scene. Users have two options: They can either select a previously saved instance of the scene on the canvas and load it by pressing the physics-reactive button, or if no saved instance is selected and the button is pressed, the simulation starts from a fresh state with no preloaded model. Upon loading, users are initially presented with a floating panel, an empty plane, and a button pillar that allows them to return to the scene selection.

   The floating panel consists of four buttons, each corresponding to one of four basic shapes: cube, cylinder, sphere, or plane. These shapes can be manipulated in terms of scale and orientation using bimanual manipulation techniques interpreted by the MRTK, described earlier in §2.2.2. These manipulations are performed through ray-casting, whereby users align a reticle over an object and then interact by grabbing a corner for scale manipulation or the center of an edge for horizontal or vertical axis rotation. If an object is positioned too far beyond the near field to be directly grabbed by hand, users can resort to ray-casting-based interactions by pinching their fingers when a reticle beamed from one's hand collides with the object. These interaction methods are illustrated in **Figure 8**. Additionally, every object is touchable, and the haptic stylus can trace the shape of each object. However, for positional translation over long distances,

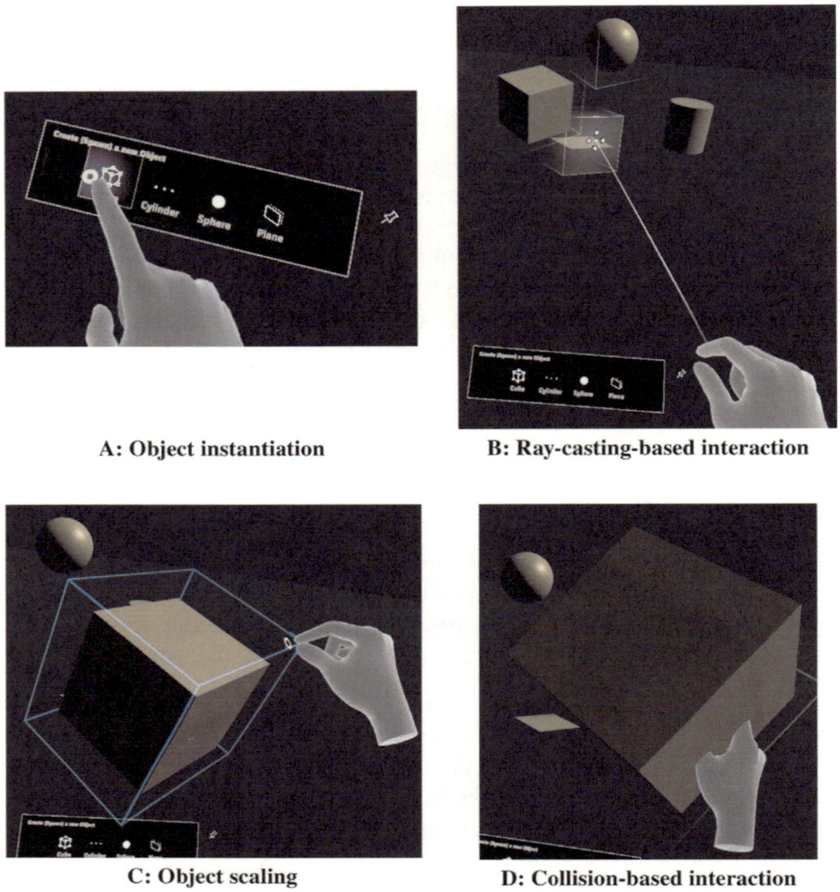

Figure 8.
Immersive haptic modeling: (A) object instantiation; (B) ray-casting-based interaction; (C) object scaling; and (D) collision-based interaction.

users must be proximate to the object, as the stylus's mechanical limitations restrict its arm from achieving large positional movements.

In addition to the objects, a plane located in front of the button pillar can be shaped using the haptic stylus. When users teleport close to the plane, they can place the stylus on top of it, press the primary selection button, and directly manipulate its surface, as shown in **Figure 9**.

3. Experimental validation

3.1 Pilot experiment

In this section, we outline two conditions under which participants in a subjective experiment experienced haptic feedback. The purpose of these tests was to identify shortcomings in the current implementation and to assess whether the ambulatory design enhances perceived immersion.

Figure 9.
Virtual sculpting.

3.1.1 Conditions: Seated (with desktop PC monitor) and mobile VR (with HMD)

The baseline condition involved connecting the haptic device to a standard desktop computer with a monitor. In this seated experience, participants played a game of Jenga where they could pick up and remove small blocks. These blocks were affected by physics, including simulated gravity, so pushing the virtual tower with the stylus would cause it to react accordingly, such as shifting or toppling.

In contrast, the room-scale (ambulatory) condition involved each participant donning our harness (with guidance) for the immersive experience and its various segments, as described in §2.2.3.

3.1.2 Procedure and controls

In the baseline condition, participants were initially introduced to the haptic device, including its controls and expected behavior. They were then shown a demonstration of a game of Jenga. This task had no time limit or specific quantitative objective; it aimed to familiarize each participant with the concept of using haptic feedback devices in a desktop setting.

In the room-scale condition, testers were first introduced to the combined VR headset and force-feedback stylus used in the previous segment. They were then guided through the haptic sandbox, briefly introducing each segment. Similarly, when transitioning to the sculpting and carving scene, subjects received instructions on controls and capabilities, and were given opportunity to explore its features. There were no time limits, quantitative objectives, or specific goals for these tasks.

Throughout the experiment, each participant experienced the same desktop scene on the same computer, with the same monitor, and in the same seating position. Additionally, every ambulatory tester wore the same harness, laptop, and VR headset. They had an opportunity to explore the same sections of the haptic sandbox as in the room-scale condition.

3.1.3 Participants

A total of 8 individuals took part in the pilot experiment, with ages ranging from 19 to 35. Participants comprised six males and two females. Their levels of experience with VR varied, ranging from no experience at all to owning an HMD and occasionally using it. This diversity allowed us to observe how intuitive the experience was and to determine if inexperienced users required more guidance than experienced users. Regarding haptic devices, participants had varying levels of experience, but the majority had no experience or had only tried them a few times. Regardless of prior experience, the need for guidance was similar due to the novelty of the featured system. This suggests that participants required more instruction regarding operating the haptic stylus compared to that using the headset alone. However, once basic interactions were explained, no further guidance was needed.

3.1.4 Data acquisition and composition

Following the experience with both setups, participants were given a questionnaire to complete. The questionnaire consisted of assertions related to their impressions, and participants were asked to indicate their level of agreement on a quantified Likert scale. Furthermore, participants were asked a few additional questions regarding prior experience with haptic devices, virtual reality, and CAD software.

3.1.5 Results

Participants' quantified impressions of the immersiveness of the desktop experience varied, but we were able to conclude that users generally had neutral or slightly positive feelings of immersion while playing the simulated Jenga game. Some concerns were raised about the weight of the setup, and several participants indicated a comfortable wearability time of a quarter to half an hour. Female participants specifically noted that the front of the harness should be softer by adding more padding between the vest and the aluminum plating. Overall, the response was more neutral than negative. It was encouraging to confirm that the simulation of tactile feeling was considered close to a realistic sensation. Most users agreed that the simulation provided an experience comparable to natural touch sensations. However, some users expressed disappointment with the device intensity, reporting that certain effects were not strong enough to be on par with real-life sensations.

When participants were asked about immersiveness of the VR mode, the feedback was overwhelmingly positive. Despite the weight and occasional discomfort of the harness, the sensation of immersion was not diminished. Comparing feedback on the form-factor immersiveness between the immersive and the desktop modes clearly showed that the combination of VR and haptic force-feedback contributed to overall immersion. Most users found the combination of inputs usable but somewhat tricky. A rotational reset function had been provided to help users realign the virtual stylus with its real-world affordance after teleportation. However, many participants noted that the congruence between the virtual stylus and its real-world counterpart sometimes drifted, and the rotational reset function could not be fully relied upon.

Hardware limitations of the stylus and perceived inadequate intensity of haptic effects resulted in no participant choosing "Natural" as their impression.

Overall, hand-tracking was regarded as quite accurate, but the transition between using the haptic stylus and returning to using one's dominant hand for gestures required users to hide their hand and then look at it again to resume the hand-tracking mode. Although participants considered this awkwardness as something they would "get used to," it will be addressed in future versions. All participants agreed that this type of device arrangement could be used for CAD applications. When asked to rate overall experience on a scale from 1 to 10, participants provided quite positive feedback. The quality of our proof-of-concept received an average score of 8.5 out of 10. Any score below 5 would be considered unsuccessful, so achieving "success" with our prototype was gratifying. However, there is still ample room for improvement. The feedback we received in the form of complaints, compliments, and suggestions for future refinement and expansion was invaluable [32, 36].

3.2 Performance experiment

In a subsequent experiment conducted several months after the previous one, our focus shifted from characterizing absolute performance to confirming that the ambulatory performance was at least on par with the performance measured under the fixed condition. The hardware used in the improvement experiment remained the same, and no subjects participated in both experiments.

3.2.1 Conditions: Seated (with desktop PC monitor) and mobile VR (with HMD)

The baseline condition closely resembles that discussed in §3.1.1, with the inclusion of the block-sorting game as outlined in §2.2.3. However, the Unity application was re-engineered to ensure precise correspondence with the room-scale condition. The composition of this scene is shown in **Figure 10**.

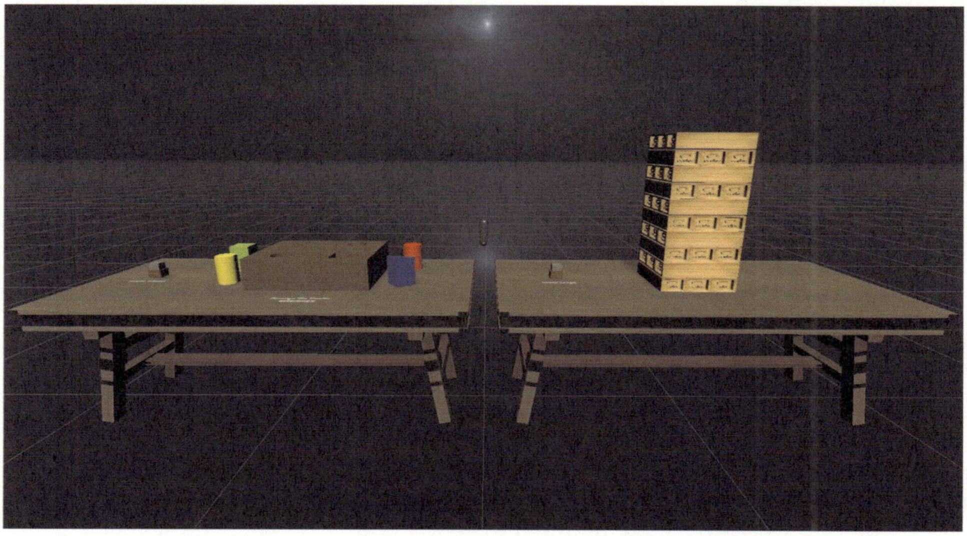

Figure 10.
Performance experiment scene: both desktop and immersive conditions.

As mentioned in §3.1.1, the room-scale setup involved equipping each participant with our harness. However, instead of allowing them to independently explore the features, they were instructed to complete a set of predetermined tasks, as described following.

3.2.2 Procedure and controls

For the Jenga game, participants were instructed to remove as many blocks as possible from the tower within a 4-minute time limit without causing it to topple. If the tower toppled, they had the option to reset and start over. The highest number of removed blocks achieved from any number of attempts was recorded, along with the number of resets.

Similarly, in the shape-sorting game, testers were allotted a 4-minute time interval to sort the complete set of shapes. The score was incremented only if all shapes were successfully sorted, preventing players from selectively sorting only easier shapes.

Participants were divided into two groups: One group experienced only the ambulatory segment, while the other group exclusively engaged in the desktop experience. This partition was implemented to avoid learning effects and biased results favoring either of the two conditions based on a tester's increased experience through the experiment.

The measured segment lasted approximately 10 minutes, excluding introduction of the experiment to each participant, warm-up session, and questionnaire completion. The warm-up session took about 2 minutes for each segment (4 minutes in total), and answering the questionnaire required up to 10 minutes. Overall experiment experience duration ranged from 20 to 30 minutes per participant.

In the desktop segment, each participant used the same computer, monitor, and maintained the same seating position (with only seat height adjusted to align the monitor with each tester's eye level).

During the warm-up period of the ambulatory segment, our focus was primarily on adjusting the harness to ensure participant comfort and prevent any discomfort that could potentially affect results. Additionally, it provided an opportunity for the subject to become familiar with the new interface.

3.2.3 Participants

A total of eight adult participants volunteered for the ambulatory experiment. Among them, 3 (37.5%) were aged between 18 and 25, while the remaining 5 (62.5%) were aged between 26 and 35. In terms of gender distribution, 6 (75%) were males and 2 (25%) were females.

Similarly, eight adults took part in the desktop version of the experiment. Among them, 7 (87.5%) were aged between 18 and 25, and 1 (12.5%) was aged between 26 and 35. In this group, 5 (62.5%) were males and 3 (37.5%) were females.

All participants received compensation for their participation, receiving ¥1000 (approximately $8) for a half-hour session. All participants were right-handed and had a background in Computer Science or Software Engineering, making them well-versed in standard human-computer interaction practices. However, some participants had no previous experience with VR. They were introduced to the basic concepts and usage of VR by allowing them to walk in Meta Home, and they were shown how to enable the pass-through "Chaperone" functionality of the Quest 2 HMD (a.k.a "Guardian" for Oculus systems), which provides an optical representation of one's physical environment using camera capture and video see-through, reassuring them about the minimal risk of accidental collision with real objects.

3.2.4 Data acquisition and composition

Data was collected by the experimental supervisor through direct observation using a stopwatch and Google Forms to record scores. In addition to the recorded data, each subject was asked to complete a User Experience Questionnaire (UEQ) after each measured segment. The UEQ consisted of 26 pairs of bipolar dichotomies, such as "complicated–easy" and "inventive–conventional." Participants indicated their evaluation of the User Experience (UE) on a quantized Likert scale ranging from 1 to 7 [41, 42].

Furthermore, subjects were also asked to respond to 36 questions selected from the multidimensional scale Intrinsic Motivation Inventory (IMI). The IMI statements, such as "I was pretty skilled at this activity" and "This activity was fun to do," were contradicted or confirmed by indicating level of agreement on a 7-step scale ranging from "not true at all" to "very true" [43].

The combination of observed data, scores, UEQ responses, and IMI assessments provided comprehensive overview of the participants' experiences and subjective evaluations.

3.2.5 Results

Results of the UEQ were analyzed and categorized into six dimensions: Attractiveness, Dependability, Efficiency, Novelty, Perspicuity, and Stimulation. Scores were derived from a zero-centered seven-point scale (−3 to +3). Statistical analysis using the ANOVA method with the 'ez' library in R revealed no significant differences between the desktop and ambulatory versions of our application. However, when benchmarked against data from 21,175 individuals in 468 studies on various products, the desktop segment scored slightly lower than the ambulatory segment. Detailed results can be found in **Table 1A** and **Figure 11**.

The IMI was used to assess intrinsic motivation across six subscales: Effort and Importance, Interest and Enjoyment, Perceived Choice, Perceived Competence, Pressure and Tension, and Value and Usefulness. ANOVA analysis of the experimental results indicated no significant differences between the desktop and ambulatory conditions. Results are presented in **Table 1B** and **Figure 12**.

(A) UEQ Results ($p < 0.05$)		**(B) IMI Results ($p < 0.05$)**	
Dimension	**Statistics**	**Dimension**	**Statistics**
Attractiveness	$F(1, 14) = 0.3378; p = 0.5703$	Effort and Importance	$F(1, 14) = 0.0029; p = 0.9574$
Dependability	$F(1, 14) = 0.0068; p = 0.9356$	Interest and Enjoyment	$F(1, 14) = 0.0204; p = 0.8886$
Efficiency	$F(1, 14) = 0.0333; p = 0.8578$	Perceived Choice	$F(1, 14) = 0.1639; p = 0.6917$
Novelty	$F(1, 14) = 0.0000; p = 1.0000$	Perceived Competence	$F(1, 14) = 1.4299; p = 0.2516$
Perspicuity	$F(1, 14) = 0.7221; p = 0.4098$	Pressure and Tension	$F(1, 14) = 1.4246; p = 0.2525$
Stimulation	$F(1, 14) = 0.2355; p = 0.6350$	Value and Usefulness	$F(1, 14) = 0.2056; p = 0.6572$

(C) Performance results ($Pr < 0.05$)	
Independent variable	**Statistics**
Max. number of removed and stacked Jenga blocks	$\chi^2 = 0.1763$; Pr $(>\chi^2) = 0.6746$
Number of Jenga trials (resets +1)	$\chi^2 = 0.0044$; Pr $(>\chi^2) = 0.9471$
Number of filled shape sorter boards	$\chi^2 = 1.9884$; Pr $(>\chi^2) = 0.1585$

Table 1.
*Experiment results: **(A)** user experience questionnaire; **(B)** intrinsic motivation inventory; and **(C)** performance.*

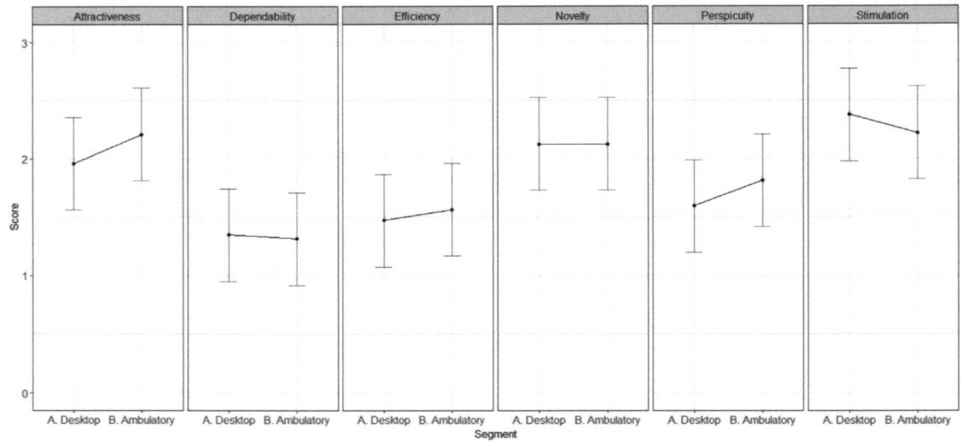

Figure 11.
Performance experiment (desktop and ambulatory conditions)—Compiled UEQ results of zero-centered seven-point scale with ordinate axis truncated to positive interval [0,3]. Error bars correspond to confidence interval of 95%.

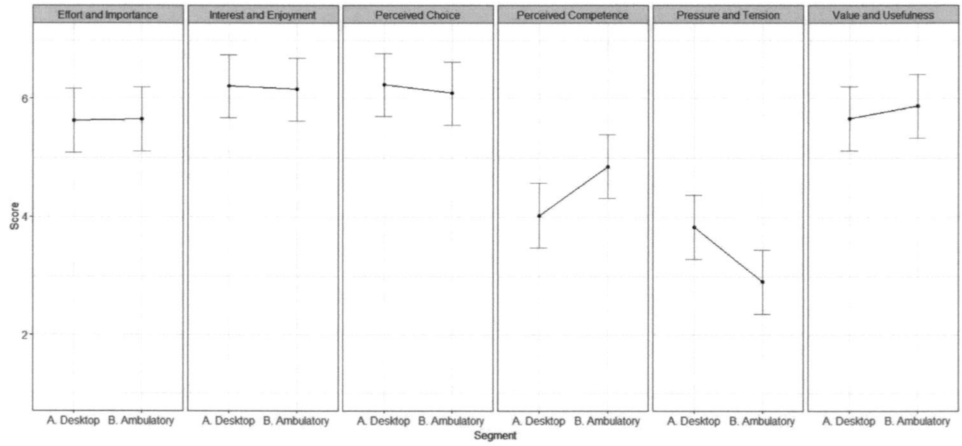

Figure 12.
Performance experiment (desktop and ambulatory conditions)—IMI results; error bars correspond to confidence interval of 95%. A single irrelevant question and an irrelevant scale from IMI were excluded from the analysis.

Performance measurements were analyzed using the ANOVA function in R and the Type II Wald χ^2 test for logistic regression analysis. Three performance metrics were considered: the maximum number of removed and stacked Jenga blocks, the number of Jenga trials (including resets), and the number of filled shape sorter boards. Analysis confirmed that the performance measurements, shown in **Table 1C**, did not exhibit any significant differences between the two conditions, as indicated by $Pr(>0.1763) = 0.6746$, $Pr(>0.0044) = 0.9471$, and $Pr(>1.9884) = 0.1585$, respectively. Since each Pr value is greater than the rejection threshold of 0.05, we can conclude that there were no significant differences in performance between the two conditions [32, 36].

3.3 Implemented enhancements and extensions

Based on the experiments summarized in §3.1.5 and §3.2.5, several limitations were identified that negatively impact user experience with the prototype. These observations provide valuable insights into areas that require improvement. Despite the haptic feedback enhancing immersiveness in this specific use case, it did not improve performance. This indicates that enhancing performance and user experience could potentially further enhance the sense of immersion and depth of presence. Addressing these concerns would involve improving comfort, enhancing realism of touch sensations, refining input usability, ensuring alignment accuracy, and facilitating smoother transitions between different interaction modes within the system experience.

3.3.1 Reducing weight

Improvement efforts focused on enhancing comfort aspects of the prototype and addressing software-related issues mentioned earlier. Weight reduction was a key objective during this phase. The main goal was to minimize the number of components attached to the harness. We reduced the weight significantly by relocating the laptop computer and the rear support plate, which were originally positioned at the back of the user. Our specific use case required maintaining untethered six degrees-of-freedom movement within the virtual space. Therefore, we needed to shift from wired to wireless connectivity between the Quest 2 HMD and 3D Systems Touch haptic interface.

3.3.2 Establishing wireless connectivity

The transition from Quest Link to Air Link, where an HMD acts as a thin client for PC, was a relatively straightforward process. However, it required upgrading our development environment, Unity editor, and its Oculus XR libraries to newer versions to ensure smooth and reliable functionality with Air Link. Unthetering the haptic device from wired to wireless architecture presented significant challenges. Initially, we explored the option of converting the connection from USB to Bluetooth at the hardware level, but ultimately decided to use USB over a wireless network. To achieve this, we utilized a VirtualHere server running on a Raspberry Pi 3, equipped with a Wi-Fi adapter capable of 5 GHz connection. VirtualHere allows the network to serve as a conduit for transmitting USB signals, effectively allowing USB over IP. This USB server solution is best for distributed deployment of USB devices over a local area network (LAN), without the need for wireline physical connection to a client machine. The USB device behaves as if it were directly connected to, in this case, a laptop computer, even though it is physically plugged into a remote server (Raspberry Pi 3). Consequently, existing drivers and software function seamlessly without requiring special modifications.

However, there is a slight latency that slightly impacts our implementation under certain conditions. As explained in §2.2.2, Unity utilizes the `OHToUnityBridge.dll` library, while vanilla applications from 3D Systems directly interface with the device through HD/HL framework. When the stylus is driven directly using HD/HL libraries, the latency and reliability of haptic feedback are indistinguishable from when the device is physically connected to the target machine. However, when using an application built in Unity, the additional runtime overhead of the translation layer, which involves converting native API calls through middleware to express physical properties of virtual objects using the stylus, introduces occasional noticeable delays in force-feedback response and subtle stylus jitter. Occurrence of these issues depends upon various factors, such as the environment and Wi-Fi signal quality, which may be beyond our control. Furthermore, even without perceptible issues, a slight degradation in the servoloop frequency, which facilitates bidirectional communication between the application and the device, can be observed when comparing wired and wireless communication. We plan to address these concerns and enhance the interface in future updates by exploring ways to minimize middleware overhead experience.

3.3.3 Power delivery

Deployment and management of wireless communication between the host laptop, client HMD, and stylus device allows for reduction in power delivery requirements for all devices involved. Previously, we were limited by the built-in battery of our host machine, which had a capacity of 90 Wh. Through various power-saving strategies, such as CPU undervolting and throttling clock speeds of the CPU and GPU, we could achieve a screen-on time of approximately 60 to 90 minutes. However, by eliminating the need for the user to carry a laptop computer, we can disable all power-saving measures, improve rendering performance, and achieve higher texture quality and visuals within the limitations of the Air Link function of Meta Quest 2.

The power delivery for the stylus device as described in §2.2.1 remains unchanged, utilizing an external power bank. However, the power bank now needs to supply power to the Raspberry Pi 3 as well, which reduces the stylus device's power-on time. Previously estimated at approximately 12 hours, system-on time can now be estimated to be a maximum of about 10 hours. Since a 10-hour run-time of this system exceeds presumed continuous session duration for a single user, we decided to utilize the same external power source to extend the battery life of the HMD as well. Considering the average power consumption of the Quest 2 (4.7–7 W, Raspberry Pi 3 (1.3–3.7 W), and the 3D Systems Touch (18–31.5 W), the entire system can be expected to run on a single charge of a 20 Ah battery for approximately 4.5 to 8 hours, effectively quadrupling minimum system up-time. The improved usability time resulting from the weight reduction and ergonomic enhancements described in §3.3.1 only extends to 2 hours. Therefore, power requirements do not present a significant limitation for this system. All the hardware changes mentioned above and the overall hardware assembly are illustrated in **Figure 13**.

3.3.4 Enhancing haptic feedback intensity

As previously mentioned, there were concerns regarding perceived intensity of certain haptic effects. To address this issue, specific adjustments were implemented to improve simulation of physical properties for virtual objects. The simulation process is

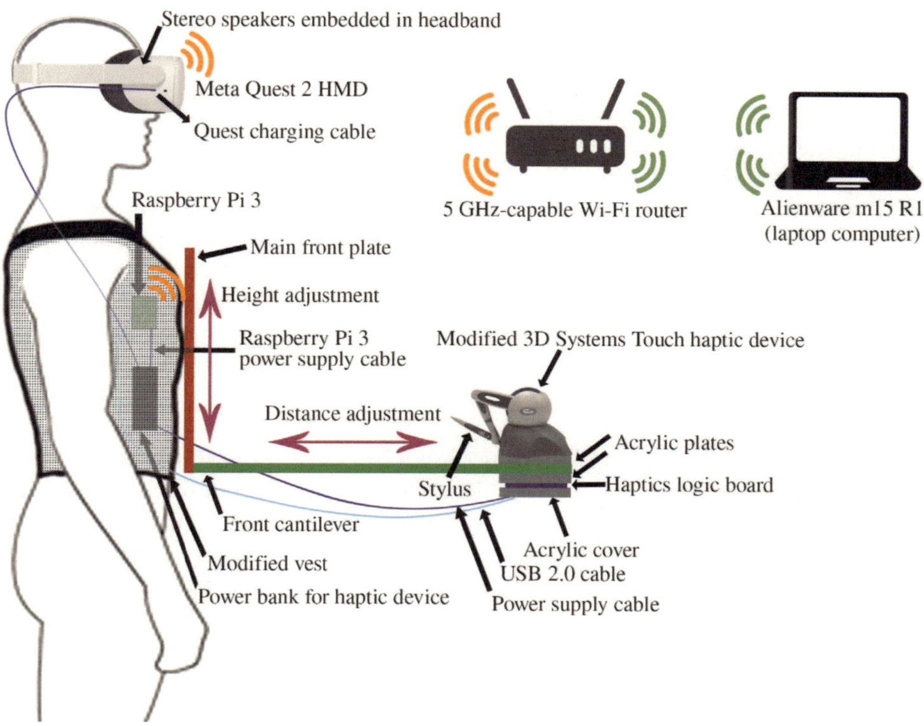

Figure 13.
Wireless system assembly diagram—"elevation" (profile).

handled by Unity's physics engine, which compiles and renders simulated properties into forces exerted by the stylus. To create more realistic experience when interacting with virtual objects using the stylus, various properties were reviewed and modified.

These properties encompass elements such as the perceived weight of objects, their drag, bounciness, friction coefficients of different materials, smoothness, stiffness, and intensity of damping when the stylus comes into contact with an object. By fine-tuning these properties, our aim has been to enhance the immersive nature of the tactile experience, making it closely resemble real-life interaction. These adjustments enable users to perceive and interact with virtual objects in a manner that aligns with expectations, ultimately providing a more satisfying and engaging haptic and overall experience.

3.3.5 Alignment and reset function for stylus to avatar

Previously, there were challenges in maintaining consistent alignment and synchronization between the user's virtual avatar and the stylus device. When utilizing locomotion features, such as hand-gesture-initiated teleportation, slight drift or rotational misalignment could occur between the virtual representation of the stylus and its physical counterpart. In the implementation described in §2.2.2, the MRTK was utilized to enable hand-tracking and gesture-operated teleportation. However, teleportation caused the avatar to independently rotate around its gravitation vertical axis (yaw), separate from the orientation of the stylus attached to the avatar. To address this discrepancy, a reset function was introduced, allowing users to realign the stylus and avatar (by simultaneously pressing two buttons on the stylus).

To accommodate the switch to the latest version of the Oculus XR Plugin, departure from the outdated MRTK was necessary. This switch involved reintegration of features previously provided by MRTK into the application. The newer XR Interaction Toolkit was employed for this purpose, aiding in the implementation of the updated locomotion system. From the user's perspective, the locomotion function operates in a seamless manner, while resolving the issue of independent rotation between the avatar and stylus after each teleportation. The updated system ensures that the user and their stylus face the same direction as prior to initiating each teleportation event.

3.3.6 Transition between stylus use and hand-tracking

The integration of the XR Interaction Toolkit in Unity applications improved hand-tracking and controller tracking, addressing the problem of inconsistent transitions between stylus and hand-tracking for the user's dominant hand. These adjustments resulted in smoother tracking accuracy and more natural hand gestures. All the aforementioned software changes and the overall software architecture are illustrated in **Figure 14**.

3.3.7 Mixed reality pass-through

The Meta Quest 2's Mixed Reality (MR) pass-through feature offers an advanced capability that enables users to integrate their real physical environment into a virtual reality experience. Utilizing built-in cameras of the

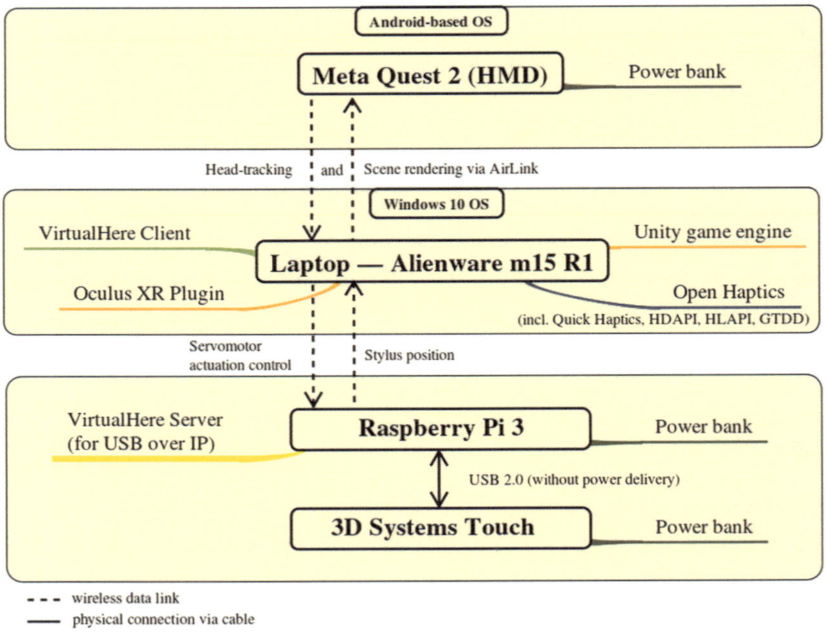

Figure 14.
Wireless software architecture.

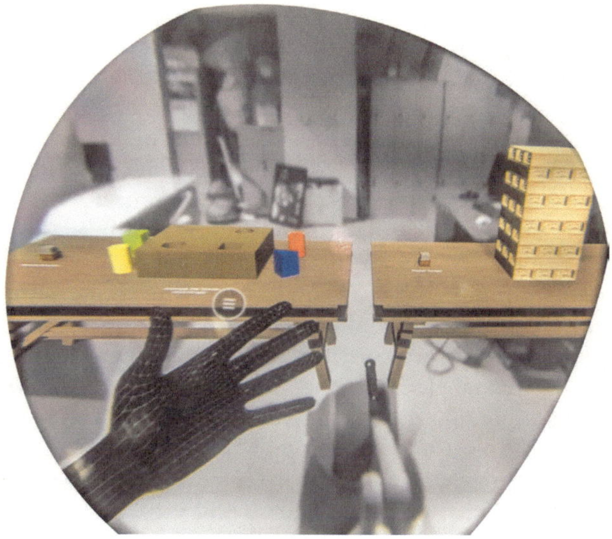

Figure 15.
The MR scene (monocular left half of binocular view) showcases virtual shape sorter and Jenga desks overlaid upon photographic pass-through imagery rendered as a skybox.

headset, this feature captures live stereoscopic video of the real world and composites it around the virtual environment. This integration allows users to perceive and interact with ambient surroundings while wearing the VR headset. As a result, virtual objects can occupy the user's physical environment, providing a blended virtual and real experience.

The MR pass-through feature not only allows users to see their real surroundings but also enables them to navigate their physical space, avoid obstacles, and interact with real objects while immersed in the virtual world. It enhances situational awareness, enhancing user safety and reducing collisions with physical objects. Furthermore, it enables users to incorporate real-world elements into virtual experiences for augmented reality effects.

When combined with haptic force-feedback devices like our prototype or similar devices mentioned in §1.4, the MR pass-through feature of the Meta Quest 2 goes beyond traditional VR experiences, offering a multimodal encounter that merges virtual and physical realms. This combination provides users with heightened sensory stimulation, allowing them to enjoy the realism and interactivity of a virtual environment while remaining fully aware of and engaged with physical surroundings. The result is a captivating and immersive experience that extends the boundaries of what is possible in the realm of purely virtual reality.

Figure 15 shows a monocular view of the MR pass-through from within the XR environment.

4. Future work

Currently, the haptic stylus's base is not explicitly tracked; instead, it is positioned based on the user's height and arm's length. The distance from the user's torso (X-axis) and chin (Y-axis) parameterizes the Unity scene, and the virtual representation of the stylus is offset from the anchor point of the headset. As a result, the position of the stylus's base is referenced by the HMD's position within the scene and is confined to the play-space limited by the Quest 2's Guardian mechanism. However, this setup poses a limitation. When a user leans sideways without moving their hips, the virtual stylus moves alongside this movement, while the real stylus remains in the same place, creating a tracking disconnect. To address this limitation in the future, an additional pair of cameras for image or object recognition could be utilized. By incorporating technologies such as OpenCV or other image processing frameworks, we could analyze the real space surrounding the user and estimate the true position of the haptic interface. This improvement would enhance overall tracking accuracy and provide more realistic experience for users. Furthermore, the current system only allows tactile perception with virtual elements that are pre-made and part of the scene. Future improvements involving depth cameras or other environment-scanning technologies could enable realtime rendering of "aftermarket" real-world objects into simplified virtual representations. This extension would create elements within the realm of augmented virtuality, blending real and virtual objects. This concept has been partially tested by scanning a laboratory environment using an iPad Pro and its LiDAR sensor, followed by post-processing in Blender and import into a Unity scene. The potential outcome of such developments is the ability to immerse the user in a "portable room" regardless of actual physical location. This would open up exciting possibilities for various applications, such as remote collaboration, training simulations, and interactive experiences that combine the virtual and physical worlds.

5. Conclusions

In addressing the challenges outlined in §2.1, we developed a free-range haptic interface that enables untethered six degrees-of-freedom in virtual reality, providing small- and medium-scale force-feedback. Through the transformation of a 3D Systems Touch haptic device into an ambulatory version and development of supporting software, we achieved literally tangible results in haptic technology. This innovation allows for immersive experiences in fields such as CAD, gaming, and virtual simulations, presenting previously intangible phenomena in a palpable manner. Integration of spatially flexible force-feedback displays offers new possibilities, such as ambulatory interaction with extended springs or realistically simulating organ transplants by providing haptic force-feedback in space.

To evaluate effectiveness of our solution, we conducted two experiments. The first assessed the precision, ergonomics, stability, and usability of our hardware and software, revealing certain deficiencies. However, despite these limitations, the overall results indicated the potential of combining haptic force-feedback and ambulatory VR to enhance immersion in free-range virtual environments. The second experiment focused on user experience and performance evaluation, comparing the ambulatory setup to the traditional stationary version. No significant differences were found across measured dimensions. Considering the limitations and the identified challenges from the first experiment, the absence of significant differences should not be regarded as a negative outcome. Instead, it highlights the potential of the ambulatory setup to surpass the traditional desktop version in terms of user experience and performance as hardware and software issues are addressed.

Moreover, we made substantial improvements in the wearable force-feedback mechanism and incorporated an MR pass-through feature. These enhancements encompassed weight reduction, wireless connectivity, power delivery, haptic feedback intensity, stylus alignment, and smooth transitions between stylus use and hand-tracking. Introduction of MR pass-through has been particularly impactful, as it allows users to merge real-world environments with augmented virtual elements. This integration softens boundaries between virtual and physical realms, creating coherent multimodal experience. The prototype, refined with these advancements, holds encouraging potential for further exploration in MR applications, presenting new opportunities for interactive immersive experiences.

Acknowledgements

We express our gratitude to Julián Alberto Villegas Orozco, Camilo Arévalo Arboleda, Georgi Mola Bogdan, Isuru Jayarathne, Wen Wen, James Pinkl, and our experimental subjects for their valuable support and feedback.

Abbreviations

API	application programming interface
AR	augmented reality
CAD	computer-aided design
DLL	dynamic-link library
EMG	electromyography
HCI	human-computer interaction
HDAPI	Haptic Device API
HLAPI	Haptic Library API
HMD	head-mounted display
IMI	Intrinsic Motivation Inventory
LAN	local area network
MR	mixed reality
MRF	magnetorheological fluid
MRTK	Mixed Reality Toolkit
PGM	pneumatic gel muscle
RBF	radial basis function
SDK	software development kit
TUI	tangible user interface
UE	user experience
UEQ	User Experience Questionnaire
UI	user interface
USB	Universal Serial Bus
VE	virtual environment
VR	virtual reality
XR	extended reality

Author details

Peter Kudry* and Michael Cohen*
Spatial Media Group, University of Aizu, Tsuruga, Fukushima, Japan

*Address all correspondence to: peterkudry@protonmail.com and xilehence@gmail.com

References

[1] Link E. US1825462A - Combination training device for student aviators and entertainment apparatus. 1931. Available from: https://patents.google.com/patent/US1825462A/

[2] Kato H, Billinghurst M, Poupyrev I, Imamoto K, Tachibana K. Virtual object manipulation on a table-top AR environment. In: Proceedings IEEE and ACM International Symposium on Augmented Reality. New York City, United States: Institute for Electrical and Electronics Engineers (IEEE); 2000. pp. 111-119

[3] AdsReality A brief history of augmented reality. 2020. Available from: http://adsreality.com/history-of-augmented-reality-infographic/

[4] VR society, history of virtual reality. 2020. Available from: https://www.vrs.org.uk/virtual-reality/history.html

[5] Marr B The 5 biggest virtual and augmented reality trends in 2020 everyone should know about. 2020. Available from: https://www.forbes.com/sites/bernardmarr/2020/01/24/the-5-biggest-virtual-and-augmented-reality-trends-in-2020-everyone-should-know-about/

[6] Enterprise Alliance, A. XR Industry Insight Report 2019–2020: Featuring Oculus Rift, HTC Vive, Intel, Nvidia and More - AREA. 2020. Available from: https://thearea.org/ar-news/xr-industry-insight-report-2019-2020-featuring-oculus-rift-htc-vive-intel-nvidia-and-more/

[7] Lang B. 13 major new features added to Oculus Quest 2 since launch. 2022. Available from: https://www.roadtovr.com/oculus-quest-major-feature-updates-since-launch/3/

[8] Mathur N. What's New in VR Haptics? Livery Place, Birmingham: Packt Publishing; 2018. Available from: https://hub.packtpub.com/whats-new-in-vr-haptics/

[9] Kreimeier J, Hammer S, Friedmann D, Karg P, Bühner C, Bankel L, et al. Evaluation of different types of haptic feedback influencing the task-based presence and performance in virtual reality. In: PETRA '19: Proceedings of the 12th ACM International Conference on PErvasive Technologies Related to Assistive Environments. Vol. 19. New York, NY, United States: Association for Computing Machinery; 2019. pp. 289-298

[10] Burdea G. Force and Touch Feedback for Virtual Reality. New York, NY, United States: John Wiley & Son, Inc.; 1996

[11] Rosenberg L. A Force Feedback Programming Primer: For Gaming Peripherals Supporting DirectX 5 and I-Force 2.0. Aventura, Florida, United States: Immersion Corporation; 1997

[12] Azmandian M, Hancock M, Benko H, Ofek E, Wilson A. Haptic retargeting: Dynamic repurposing of passive haptics for enhanced virtual reality experiences. In: Proceedings in CHI Conference on Human Factors in Computing Systems. New York, NY, United States: Association for Computing Machinery; 2016. pp. 1968-1979

[13] Suzuki R, Hedayati H, Zheng C, Bohn J, Szafir D, Do E, et al. RoomShift: Room-scale dynamic haptics for VR with furniture-moving swarm robots. In: Proceedings in CHI Conference on Human Factors in Computing Systems. Vol. 20. New York, NY, United States:

Association for Computing Machinery; 2020. pp. 1-11

[14] Kang N, Lee S. A meta-analysis of recent studies on haptic feedback enhancement in immersive-augmented reality. In: ACM International Conference Proceeding Series. New York, NY, United States: Association for Computing Machinery; 2018. pp. 3-9

[15] Kreimeier J, Götzelmann T. FeelVR: Haptic exploration of virtual objects. In: ACM International Conference Proceeding Series. New York, NY, United States: Association for Computing Machinery; 2018. pp. 122-125

[16] Dangxiao W, Yuan G, Shiyi L, Yuru Z, Weiliang X, Jing X. Haptic display for virtual reality: Progress and challenges. Virtual Reality Intelligent Hardware. 2019;**1**:136

[17] Choi I, Culbertson H, Miller M, Olwal A, Follmer S. Grabity: A wearable haptic interface for simulating weight and grasping in virtual reality. In: Proceedings of the Annual ACM Symposium on User Interface Software and Technology. New York, NY, United States: Association for Computing Machinery; 2017. DOI: 10.1145/3126594.3126599

[18] Slater M. Place illusion and plausibility can lead to realistic behaviour in immersive virtual environments. Philosophical Transactions of the Royal Society B: Biological Sciences. 2009;**364**:3549-3557. Available from: https://royalsocietypublishing.org/doi/abs/10.1098/rstb.2009.0138

[19] Wang D, Guo Y, Liu S, Zhang Y, Xu W, Xiao J. Haptic display for virtual reality: Progress and challenges. Virtual Reality & Intelligent Hardware. 2019;**1**: 136-162. Available from: https://www.sciencedirect.com/science/article/pii/S2096579619300130

[20] Hashizume S, Takazawa K, Koike A, Ochiai Y. Cross-field haptics: Push-pull haptics combined with magnetic and electrostatic fields. In: ACM SIGGRAPH Posters. New York, NY, United States: Association for Computing Machinery; 2016. pp. 1-2

[21] Matsumoto K, Hashimoto T, Mizutani J, Yonahara H, Nagao R, Narumi T, et al. Magic table: Deformable props using visuo haptic redirection. In: SIGGRAPH Asia Emerging Technology. New York, NY, United States: Association for Computing Machinery; 2017. pp. 1-2

[22] Girouard A, Teather R, McClelland J. Haptic feedback with HaptoBend: Utilizing shape-change to enhance virtual reality. In: Proceedings of Symposium on Spatial User Interaction. New York, NY, United States: Association for Computing Machinery; 2017. p. 150

[23] Hwang I, Son H, Kim JAP. Enhancing music playing experience in virtual reality with mid-air haptic feedback. Vol. 1. IEEE World Haptics. 2017:213-218

[24] Choi I, Sinclair M, Ofek E, Holz C, Benko H. Demonstration of CLAW: A multifunctional handheld VR haptic controller. In: The ACM Conference on Human Factors in Computing Systems, April. New York, NY, United States: Association for Computing Machinery; 2018. pp. 1-4

[25] Whitmire E, Benko H, Holz C, Ofek E, Sinclair M. Demonstration of haptic revolver: Touch, shear, texture, and shape rendering on a VR controller. In: Proceedings in CHI Conference on

Human Factors in Computing Systems, April. New York, NY, United States: Association for Computing Machinery; 2018. pp. 1-4

[26] Tzemanaki A, Al G, Melhuish C, Dogramadzi S. Design of a wearable fingertip haptic device for remote palpation: Characterisation and interface with a virtual environment. Frontiers in Robotics and AI. 2018;5:62. Available from: https://www.frontiersin.org/articles/10.3389/frobt.2018.00062

[27] Valentini P, Biancolini M. Interactive sculpting using augmented-reality, mesh morphing, and force feedback: Force-feedback capabilities in an augmented reality environment. IEEE Consumer Electronics Magazine. 2018;7:83-90

[28] Kishishita Y, Das S, Ramirez A, Thakur C, Tadayon R, Kurita Y. Muscleblazer: Force-feedback suit for immersive experience. In: 2019 IEEE Conference on Virtual Reality and 3D User Interfaces (VR). New York City, United States: Institute for Electrical and Electronics Engineers (IEEE); 2019. pp. 1813-1818

[29] Fang C, Zhang Y, Dworman M, Harrison C. Wireality: Enabling complex tangible geometries in virtual reality with worn multi-string haptics. In: Proceedings in CHI Conference on Human Factors Computing Systems. New York, NY, United States: Association for Computing Machinery; 2020. pp. 1-10

[30] Lang B. Facebook Reveals Latest Wrist-worn Prototypes for XR Input and Haptics. 2021. Available from: https://www.roadtovr.com/facebook-reveals-latest-wrist-worn-controller-prototype-for-xr-input-haptics/

[31] Shim Y, Kim T, Lee G. QuadStretch: A forearm-wearable multi-dimensional skin stretch display for immersive VR haptic feedback. In: Proceedings in CHI Conference on Human Factors in Computing Systems. New York, NY, United States: Association for Computing Machinery; 2022. pp. 1-4. DOI: 10.1145/3491101.3519908

[32] Kudry P, Cohen M. Development of a wearable force-feedback mechanism for free-range haptic immersive experience. Frontiers in Virtual Reality. 2022;3:12. DOI: 10.3389/frvir.2022.824886

[33] Allcock P. bHaptics reveals TactGlove for VR, which it will showcase at CES 2022. 2022. Available from: https://www.notebookcheck.net/bHaptics-reveals-TactGlove-for-VR-which-it-will-showcase-at-CES-2022.589393.0.html

[34] Estes A. Facebook's new haptic glove lets you feel things in the metaverse. 2021. Available from: https://www.vox.com/recode/2021/11/17/22787191/facebook-meta-haptic-glove-metaverse

[35] Stein S. Haptic gloves for Quest 2 are a small step toward VR you can touch. 2022. Available from: https://www.cnet.com/tech/computing/haptic-gloves-for-quest-2-are-a-small-step-towards-vr-you-can-touch/

[36] Kudry P, Cohen M. Prototype of a wearable force-feedback mechanism for free-range immersive experience. In: ACM International Conference Proceeding Series. New York, NY, United States: Association for Computing Machinery; 2022. pp. 178-184. Available from: https://dl.acm.org/doi/10.1145/3538641.3561507

[37] Technologies, U. About the Oculus XR Plugin — Oculus XR Plugin — 1.9.1. 2020. Available from: https://docs.

unity3d.com/Packages/com.unity.xr.oculus@1.9/manual/index.html

[38] Microsoft MRTK2-Unity Developer Documentation - MRTK 2 — Microsoft Learn. Available from: https://learn.microsoft.com/en-us/windows/mixed-reality/mrtk-unity/mrtk2

[39] Systems, OpenHaptics Toolkit Version 3.5.0 Programmer's Guide. 2018. Available from: https://s3.amazonaws.com/dl.3dsystems.com/binaries/Sensable/OH/3.5/OpenHaptics_Toolkit_ProgrammersGuide.pdf

[40] Systems, OpenHaptics Toolkit Version 3.5.0 API Reference Guide Original Instructions. 2018. Available from: https://s3.amazonaws.com/dl.3dsystems.com/binaries/Sensable/OH/3.5/OpenHaptics_Toolkit_API_Reference_Guide.pdf

[41] Schrepp M, Hinderks A, Thomaschewski J. Construction of a benchmark for the user experience questionnaire (UEQ). International Journal of Interactive Multimedia and Artificial Intelligence. 2017;**4**:40. Available from: https://www.ijimai.org/journal/bibcite/reference/2604

[42] Schrepp M, Hinderks A, Thomaschewski J. Design and evaluation of a short version of the user experience questionnaire (UEQ-S). International Journal of Interactive Multimedia and Artificial Intelligence. 2017;**4**:103. Available from: https://www.ijimai.org/journal/bibcite/reference/2634

[43] McAuley E, Duncan T, Tammen V. Psychometric properties of the intrinsic motivation inventory in a competitive sport setting: A confirmatory factor analysis. Research Quarterly For Exercise and Sport. 1989;**60**:48-58. Available from: https://pubmed.ncbi.nlm.nih.gov/2489825/

Chapter 3

Projected Augmented Reality to Display Medical Information Directly on a Patient's Skin

Pierre Boulanger

Abstract

A patient's internal anatomy can be difficult to visualize when viewed on a monitor, head-mounted display, or even when looking at an actual patient. Combining medical images (CT, MRI, US, PET) with a physical model helps recover missing anatomical context and improves situational awareness. This chapter describes an augmented reality system capable of projecting medical image information directly onto curved targets such as the human body or a mannequin. The motion of the targets and the projector are tracked using a motion capture system so that the images are adjusted in real time to match the anatomy changes in position and orientation. The augmented information can be displayed using volume rendering for realistic visualization of the internal anatomy and 3D models from segmented images. Calibration is performed on the projector and the tracking system to obtain an accurate, common coordinate system and correct visual distortions created by the fact that the projected screen (human body) is no longer a plane. The system is easily extendable to other display technology and has many potential applications, including medical education, surgical planning, and laparoscopic surgery.

Keywords: projected augmented reality, 3D tracking, medical display, image-guided surgery, multimodal image registration

1. Introduction

It is difficult to visualize and find the structures inside the human body. Generally, imaging modalities such as CT and MR are visualized as 2D slices or 3D volumes with depth and transparency using volume rendering techniques. Unfortunately, viewing these images on a remote 2D screen in the OR without the patient as a reference leads to a loss of context, particularly during surgical procedures such as Minimally Invasive Surgery (MIS) procedures. MIS are performed through one or more small incisions, using small tubes, tiny cameras (laparoscope), and surgical instruments. Another MIS approach uses robots like the Da Vinci system [1]. The laparoscopic camera (mono or stereo) provides magnified 3D views of the surgical site and helps the surgeon operate with dexterity using laparoscopic instruments. Typically, the video from the laparoscope video is displayed on a TV screen close to the surgeon (**Figure 1a**). The problem with this approach is that the surgeon's tool motion viewed in the laparoscope video is decoupled with the natural viewing direction resulting in reduced hand-eye

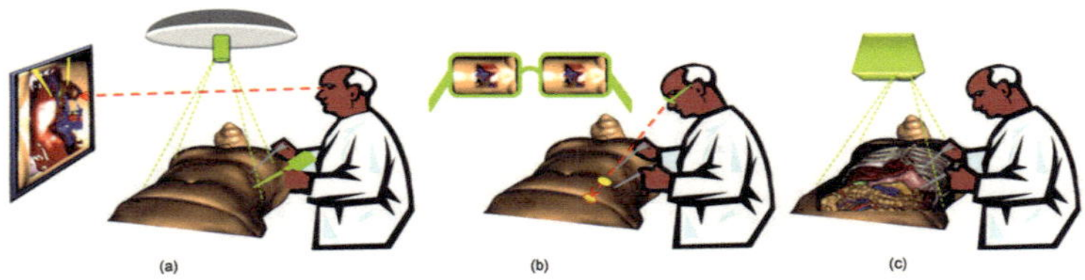

Figure 1.
Various ways to display laparoscopic video to guide MIS procedures: (a) normal screen from a video acquisition, (b) see-through display, and (c) projection display.

coordination that the surgeon must compensate for with training [2]. The second way to display a laparoscopic video is to use a head-mounted display (HMD) aligned in the same direction as the laparoscope cameras (see **Figure 1b**). This approach improves hand-eye coordination as viewing direction is now realigned with the gaze directions. A review of HMDs in surgical procedures can be found in Refs. [3, 4]. Another possibility to bring laparoscopic video into the surgeon's field of view is to project the video directly on the patient's skin using a projector. In this approach, the video projector is positioned in the OR over the patient to display the laparoscopic video on the patient skin. The result is attractive and intuitive due to a direct vision of the inner cavity resembling X-ray vision (**Figure 1c**).

This chapter describes ProjectDR, a projection-based AR system that directly displays medical information and real-time sensing on a patient's skin. In Section 1, we will review the current state of the art of Medical Augmented Reality discussing the pros and cons of very display technologies. Section 2 describes ProjectDR system configurations. Section 3 illustrates the use of ProjectDR where pre-operative models are displayed on a mannequin. We then conclude in Section 4 the pros and cons of the current system and future work.

2. Augmented reality in medicine

In the previous section, we briefly discussed three ways to display real-time laparoscopic video to improve surgeon capabilities to perform MIS procedures. In addition to real-time live data, adding pre-operative information to guide the surgeon to follow patient-specific surgical planning is possible. The combination of video and virtual augmentation is defined as Medical Augmented Reality (MAR).

Since the early experiments of Gupta *et al.* [5] in the 1990s, interest in MAR has increased substantially. In the medical context, MAR can help surgeons view patient-specific 3D models created by merging different medical imaging modalities with the patient anatomy or video with the natural gaze direction. Recent advances in graphics processors, optics, and photonics have led to the development of new low-cost commercial AR head-mounted display (HMD) systems. These devices can augment registered pre-operative and intraoperative medical imaging data to real-world images from a self-centered perspective. Pratt *et al.* [6] used Microsoft HoloLens to support reconstructive surgeries of vascular flaps. Diaz et *al.* [7] used Google Glass to perform intraoperative neuro-navigation and tumor resection. However, recent work by Cutolo [8] has shown that most consumer devices have technological and human limitations that make their use in healthcare difficult. These limitations differ depending on whether the AR technology is video-transparent (VST) or optical-transparent (OST). VST HMDs digitize the real world using one or two cameras mounted on the HMD and then present those images to the user with the registered virtual augmentations.

On the other hand, OST devices are based on optical see-through optical devices such as waveguides or semi-reflective mirrors that preserve the direct view of the world and simultaneously add computer-generated images into the user's eyes. VST systems are simpler and easier to use than OST as it is much simpler to align virtual augmentation to the video stream. In OST HMD, the registration can easily be compromised as the relationship between the eye and the lens can change. In recent OST devices, eye-tracking sensing has been used to solve the problem. Because most general-purpose HMD device focal plane is usually between 2 meters to infinity, dealing with manual tasks produces perceptual difficulties such as vergence-accommodation conflict and focus rivalry [9, 10] resulting in visual fatigue and poor hand-eye coordination. Recent work by Gabbard *et al.* [11] shows that these human-factor limitations reduce users' performance in tasks requiring a simultaneous focus on real and virtual content. To date, OST HMDs are the preferred display devices in medical AR research as they preserve a direct view of the world. However, perceptual limitations still hinder their use for high-precision talks, as described in Ref. [12]. Ferrari *et al.* [13] argued that commercial HMDs are not recommended for surgical procedures when high precision is required. In a review study on AR in Oral and craniomaxillofacial surgeries, the authors claimed that a 1–2 mm accuracy is an acceptable range [14]. In their work, Carbone *et al.* [15] suggest developing an AR HMD specifically designed for surgical guidance to meet these requirements by considering the surgical working distance of around a few tens of centimeters and correctly focusing both real and augmented information.

One solution to this problem is to develop an AR system to project virtual information directly onto the region of interest. This approach could overcome the perceptual limitations mentioned previously. Mewes *et al.* [16] use an ultra-long-throw projector to guide the radiologist during interventional magnetic resonance imaging procedures. In other work, a small LED projector in combination with a laparoscopic ultrasound has been proposed to improve efficacy and safety in laparoscopic partial nephrectomy [17]. Projected AR relies on a different paradigm than egocentric AR based on HMDs and has the potential to overcome HMD limitations [18]. However, projected AR also has some limitations. For example, parallax errors cause the location of a projected internal structure to be perceptually consistent with a single viewpoint [19]. If the surgeon's head is not tracked, as in most cases, the viewpoint coincides with the projector not the surgeon's gaze. Observing the projected AR structure from other views will produce an error in its perceived 3D position. Using the body as a projection screen bridges the gap between seeing and feeling so the user can touch the objects they see under the surface. The internal anatomy captured by the medical images can be seen in its location. With this approach, no hardware is in the way, leaving the hands free for normal interaction with tools.

3. System configuration

The required computer hardware used for this project includes a tracking system, a projector to display the images, and a computer to run the software. The tracking used was the OptiTrack motion capture system from NaturalPoint (https://www.

naturalpoint.com/). This system tracks reflective markers attached to the targets using multiple infrared cameras. The reflective markers for a target are organized into rigid bodies, where the positions of the markers are fixed relative to each other and must be visible by at least two cameras. More cameras and good positioning will result in better tracking accuracy and reduce occlusion problems. Surrounding the working area with cameras allows a full range of motion and orientation to be tracked. The required computer hardware used for this system includes a 3D tracking system for global positioning, a range sensor to digitize the skin's curved surface, a projector to display the images, and a computer to track, process, and render images. One can see in **Figure 2** ProjectDR hardware configuration.

3.1 Projector

The projector should be positioned to shine onto the desired working area. The projector can be stationary, but it is also possible to track its position with markers or to fix cameras mounted on the projector [19], allowing it to move during use. A custom mount was built so the projector could be moved by hand over a table. An Epson PowerLite 1771w projector was used to generate the image. This projector is lens-based and suffers from defocusing on very curved surfaces. This is normal, as lens-based projectors are designed to project images on planar surfaces. One can improve the defocusing effect by ensuring the projector is located directly over the region of interest by mounting it onto a gig where angles can be adjusted before the procedure. Another way to solve the problem is to use laser scanning projectors with combined RGB laser beams scanned on the skin surface using a MEMS mirror device. Because a scanning laser beam forms the image, no defocusing is present. The Nebra AnyBeam (https://www.nebra.com/) was used with success in one of our prototypes. The image quality was excellent, and no apparent defocusing was observed, even on very curved surfaces. The only issue was that many of these pico projectors suffered from a low brightness level of 150 lumens, which may not always be compatible with OR conditions.

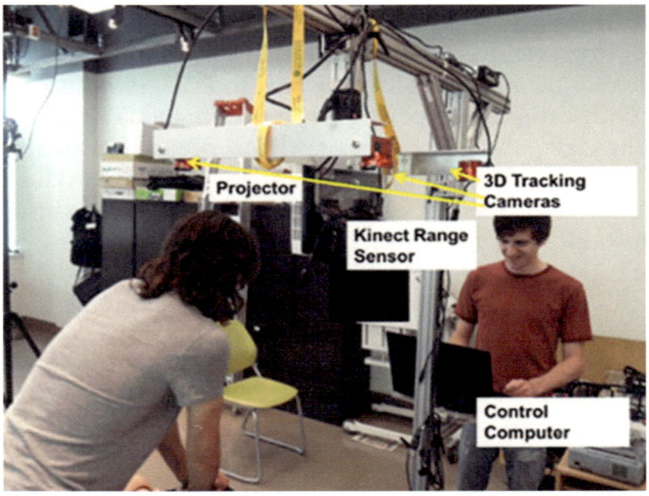

Figure 2.
ProjectDR hardware configuration.

3.2 Global tracking

Global patient tracking is performed by an OptiTrack motion capture system from NaturalPoint with a precision of 0.2 mm. Our system uses six infrared cameras to track reflective markers attached to the patient's skin. The reflective markers are organized into rigid bodies, where the positions of the markers are fixed relative to each other and must be visible by at least two cameras. Surrounding the working area with cameras allows a full range of motion and orientation to be tracked and eventually to tack surgical tools and hand motions.

3.3 Local range sensing

Besides the global patient tracking system, a Kinect range sensor captures the shape of the skin region where the information is projected. The Kinect v2 range sensor has a field of view (FOV) of 70° × 60° and an operating range from 0.5 to 4.5 m. The ground sample distance (GSD) for the Kinect is 1.4 mm in the 0.5 m range and 12 mm at 4.5 m, which is sufficient for our application.

3.4 Registering the virtual model to patient's markers

The augmentation model must first be registered to the patient's markers. To do so, retroreflective markers are installed on the patient and measured using the Optitrack system. Corresponding markers must also be placed on the virtual model. Using an Iterative Closest Point algorithm [20], scaling and the rigid transformation matrix are computed to transform the virtual model in the same scale and coordinate system as the patient.

3.5 Augmented image generation

Virtual augmentations are generated by performing a real-time ray-traced of the pre-operative model. This is achieved using an NVIDIA GeForce RTX™ 30 Series GPU capable of real-time ray-traced rendering. The graphic card is powered by Ampere—NVIDIA's 2nd gen RTX architecture with dedicated 2nd gen RT Cores. The GPU is used to ray-trace the pre-operative model from the surgeon's viewpoint captured by a 3D tracker mounter on his head. A real-time geometric image correction is applied to the rendering to compensate for the non-planarity of the patient's skin. A geometric warp function is applied to the rendered image so that the projected image appears geometrically correct for a given viewpoint on the patient's skin. The algorithm is based on the one described in Ref. [21]. The algorithm used the point cloud measured by the Kinect and the projector's intrinsic and extrinsic calibration parameters to generate this geometric warping function.

3.6 Software

The purpose of the ProjectDR software is to render different perspectives of the scene and provide control over it. ProjectDR is written in C++ with an interface in Qt and QML. The positions of the markers on the targets are streamed to ProjectDR using the NaturalPoint Motive software and NatNet SDK. The movement of the markers corresponds to the movement of any associated models in the scene and is displayed by the projector. In addition, image distortion correction is applied to the image using the information provided by the Kinect range sensor. OpenSceneGraph was used to load and render models, supporting many common model types while allowing direct control. For volume rendering, custom openGL code was written to work with OSG. This will enable ProjectDR to display volumes and polygonal models simultaneously. The software provides three main views. The "scene view" provides a zoomed-out view of all 3D models in the scene and the projector. The "model view" is for viewing and editing the position and orientation of individual models. Additional controls exist for creating a transfer function and viewing individual slices for volume rendering. The "projector view" shows what is projected onto the patient based on what a virtual camera sees in the scene from the same position and orientation as the projector. The motion tracking and projector require individual calibration, but all components must use the same coordinate system so that the virtual objects mesh with real-world targets accurately. The OptiTrack has a coordinate system that corresponds closely to the real world, can be calibrated very accurately, and is used to calibrate the other systems.

3.7 Calibration

The motion tracking and projector require individual calibration. Still, all components must use the same coordinate so that the coordinates of the virtual objects with real-world targets are accurately registered.

3.7.1 Calibrating the motion capture system

The OptiTrack system has a coordinate system, which corresponds closely to the real world and can be calibrated very accurately, so it will be used to calibrate all the other systems. The OptiTrack uses a fixed-size calibration wand that is moved around the area viewed by its six cameras to calibrate the six cameras positions, orientations, focal lengths, and lens distorsions. A right triangle with reflective markers is used to set as a reference ground plane and to set each axis's origin and direction. The objects in ProjectDR will use the same coordinates as the OptiTrack, so the origin and offsets of the models are the same as in the OptiTrack. For example, a reflective marker placed 10 cm away from the origin will appear as a dot 10 cm away from the origin in the same direction in the virtual scene.

3.7.2 Calibrating the projector

A projector has extrinsic and intrinsic parameters to calibrate as well. The extrinsic parameters describe the position and orientation of the projector relative to the origin. This must be calculated accurately since any error will result in the models not corresponding to their physical targets. Intrinsic parameters are a mapping between the pixels of the images and what is displayed by the projector. These are the smaller distortions caused by the projector and will vary between projectors. Both sets of parameters are calculated using the OptiTrack system. A small dot displayed by the projector is effectively a vector originating at the lens and extending through the area we want to use. By projecting a series of dots on a grid and placing one of the optitrack markers in its path, we can record a series of positions in 3D space for use in a standard projector calibration algorithm (cite) to calculate the extrinsic and intrinsic parameters.

To verify the accuracy of the calibration, a virtual model of a one-by-one meter grid with 10 cm squares can be projected onto the ground plane. It was measured with a ruler and should appear on the ground with the same measurements and no visible distortions if well-calibrated. The accuracy was estimated to be 0.2 mm The projector's location and orientation will now appear in the scene at the same location as in the real world.

3.7.3 Model registration

When a pre-operative 3D model is loaded into ProjectDR, it will be placed at the scene's origin point. Since that is the same location as the origin for the motion tracker, its physical target can be placed at the origin and then attached to a model through ProjectDR. The target's reflective markers are visible in the software, so the model can be moved to match the markers exactly using the model view. A model captured directly from a patient should be a close match, but scaling the models can generalize the data to work for other targets.

4. Experimental results

We used medical images from the OsiriX DICOM Image Library and a set of 3D models created from a segmented CT scan to demonstrate the system functionality. The system comprises a single computer with an Intel I7-4770 and an NVIDIA GeForce RTX™ 30 running the tracking and projection. The global motion tracking uses 12 Flex 6 cameras running at 120hz in a wide ring around the targets. An Epson PowerLite 1771w projector was chosen for its large depth-of-field and high illumination (3000 lumens). The first test was to display the vertebrae in the back (see **Figure 3a**) on a mannequin with three targets. This system can teach clinicians the anatomy of the

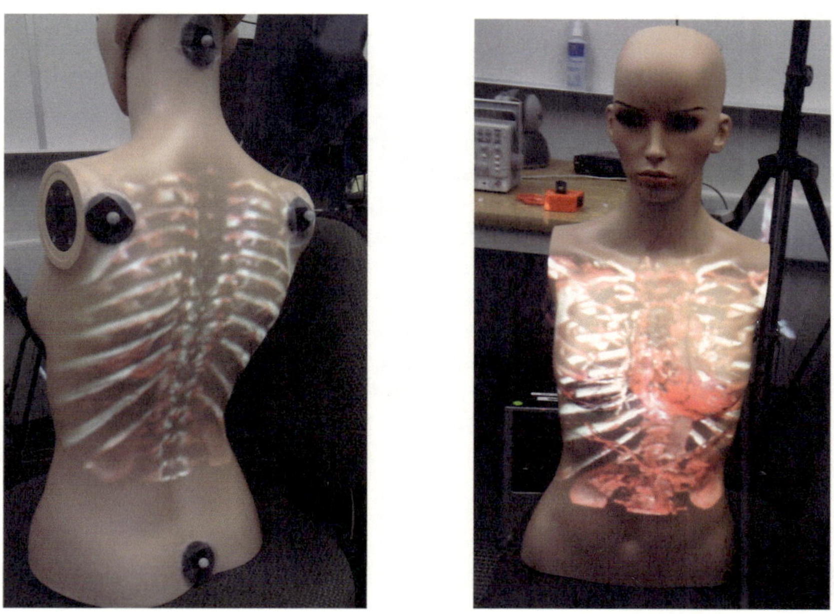

(a) (b)

Figure 3.
Project CT data rendering on a moving test mannequin: (a) back view and (b) front view.

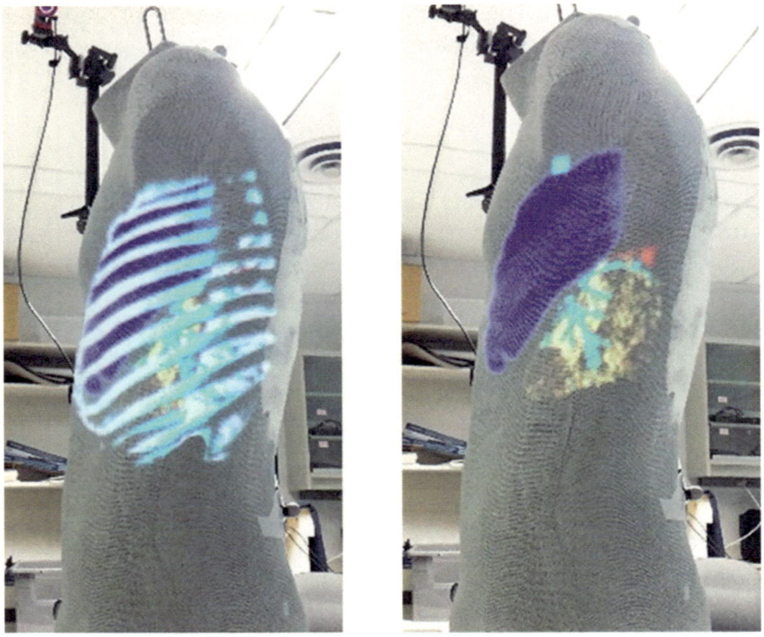

(a) (b)

Figure 4.
Projection of 3D models of the rib cage, sections of the lungs, bronchus, and a tumor (a) Rib cage is obscuring the tumor (b) Rib cage is hidden to show location of tumor, marked in red.

spine and the anatomy surrounding it. A clinician intending to palpate the spine relies upon their touch and knowledge of physical landmarks to locate a vertebra obscured by the skin. With ProjectDR, the CT image of the patient's spine can be displayed on the front (**Figure 3b**). A CT image of the thorax spine was used with a hand-crafted transfer function to show bones and some internal organs. Another application is for surgical planning. Knowing a patient's specific anatomy is essential while planning a surgical procedure. Presenting pre-operative images of the patient gives the surgeons a greater context of the task at hand. Another application is for surgical planning. It is imperative to know a patient's specific anatomy while planning an operation. Presenting pre-operative images of the patient gives the surgeons greater context. Many 3D models were used simultaneously for this example. The pre-operative models were segmented so it is possible to display each piece of the anatomy together or individually and move or hide them. **Figure 4a** shows a tumor near the heart and other important organs. Using ProjectDR, it is possible to suppress the rib cage and part of the lung, obscuring the tumor while still having a clear view of the nearby bronchus and veins (see **Figure 4b**).

5. Conclusions

This chapter describes an early prototype of the ProjectDR system that can project medical images onto a subject in real time and adapt them dynamically to changes in the patient's position and orientation. This technology has many potential applications for surgical planning, image-guided surgeries, and medical education. One of the limitations of projection-based systems is that the perspective of the images is only correct when viewed from the point of view of the projector. For example, if a heart is projected directly onto the chest, someone looking from the side would not see the heart in the correct location as one would from the front. This could be corrected by tracking the head of the user and projecting based on their perspective. Occlusion is when an object that is not being tracked moves between the projector and the targets resulting in a shadow cast across the target or projection onto the wrong object. Occlusion is a problem for all AR systems, and research has been done with HMD that could also be applied to projection systems [22]. The color and surface texture of the targets is limiting since they will blend with the projected images. Adding a color camera to the system can correct distortions by detecting differences between the desired appearance of the images and what is visible [23]. Additionally, small projectors might not be luminous enough to project onto targets in brighter lighting conditions, but advancing projector technology could improve this [24].

Future work includes improving the system by addressing its limitations and developing new applications for the technology. The depth sensors could be used to automatically detect the initial pose and correct any errors between the models and the targets to create a more automated system. A major application would be to make models of surgical tools and track them during laparoscopic Surgery. This would show the positions of the tools inside the patient relative to anatomical structures. Collaborative environments featuring users in different locations would also be possible since virtual objects and targets could be shared between systems.

Acknowledgements

The authors would like to thank NSERC for the grand supporting this work.

Conflict of interest

The authors declare no conflict of interest.

Author details

Pierre Boulanger
Faculty of Science, Deparment of Computing Science, University of Alberta, Edmonton, Canada

*Address all correspondence to: ekleczyk@targetrwe.com

References

[1] Longmore SK, Naik G, Gargiulo GD. Laparoscopic robotic surgery: Current perspective and future directions. Robotics. 2020;**9**:42

[2] Walczak DA, Pawełczak D, Piotrowski P, Trzeciak PW, Jędrzejczyk A, Pasieka Z. Video display during laparoscopy—Where should it be placed? Wideochir Inne Tech Maloinwazyjne. 2015;**10**(1):87-91

[3] Rahman R, Wood ME, Qian L, Price CL, Johnson AA, Osgood GM. Head-mounted display use in surgery: A systematic review. Surgical Innovation. 2020;**27**(1):88-100

[4] Prendergast CJ, Ryder BA, Abodeely A, Muratore CS, Crawford GP, Luks FI. Surgical performance with head-mounted displays in laparoscopic surgery. Journal of Laparoendoscopic & Advanced Surgical Techniques. Part A. 2009;**19**(Suppl 1):S237-S240

[5] Gupta SC, Klein SA, Barker JH, Franken RJPM, Banis JC. Introduction of new technology to clinical practice: A guide for the assessment of new VR applications. Journal of Medical Virtual Reality. 1995;**1**(1):16-20

[6] Pratt P, Ives M, Lawton G, Simmons J, Radev N, Spyropoulou L, et al. Through the HoloLens looking glass: Augmented reality for extremity reconstruction surgery using 3D vascular models with perforating vessels. European Journal of Radiology. Exp. 2018;**2**(1):2

[7] Diaz R, Yoon J, Chen R, Quinones-Hinojosa A, Wharen R, Komotar R. Real-time video-streaming to surgical loupe mounted head-up display for navigated meningioma resection. Turkish Neurosurgery. 2018;**28**(4):682-688

[8] Cutolo F. Letter to the editor on 'augmented reality-based navigation for computer-assisted hip resurfacing: A proof of concept study. Annals of Biomedical Engineering. 2019;**47**(11):2151-2153

[9] Oshima K et al. SharpView: Improved clarity of defocussed content on optical see-through head-mounted displays. In: Proc. IEEE Virtual Reality (VR) Conference. Greenville, SC, USA: IEEE; 2016. pp. 253-254

[10] Hua H. Enabling focus cues in head-mounted displays. Proceedings of the IEEE. 2017;**105**(5):805-824

[11] Gabbard JL, Mehra DG, Swan JE. Effects of AR display context switching and focal distance switching on human performance. IEEE Transactions on Visualization and Computer Graphics. 2019;**25**(6):2228-2241

[12] Condino S, Carbone M, Piazza R, Ferrari M, Ferrari V. Perceptual limits of optical see-through visors for augmented reality guidance of manual tasks. IEEE Transactions on Biomedical Engineering. 2020;**67**(2):411-419

[13] Ferrari V, Carbone M, Condino S, Cutolo F. Are augmented reality headsets in surgery a dead end? Expert Review of Medical Devices. 2019;**16**(12):999-1001

[14] Badiali et al. Review on augmented reality in Oral and Cranio-maxillofacial surgery: Toward surgery-specific head-up displays. IEEE Access. 2020;**8**:59015-59028

[15] Carbone M, Piazza R, Condino S. Commercially available head-mounted

displays are unsuitable for augmented reality surgical guidance: A call for focused research for surgical applications. Surgical Innovation. 2020;**27**(3):254-255

[16] Mewes A, Heinrich F, Kägebein U, Hensen B, Wacker F, Hansen C. Projector-based augmented reality system for interventional visualization inside MRI scanners. International Journal of Medical Robotics and Computer Assisted Surgery. 2019;**15**(1):e1950

[17] Edgcumbe P, Singla R, Pratt P, Schneider C, Nguan C, Rohling R. Follow the light: Projector-based augmented reality intracorporeal system for laparoscopic surgery. Journal Medical Imaging. 2018;**5**(2):1

[18] Mamone V, Ferrari SC, Cutolo F. Projected augmented reality to drive osteotomy surgery: Implementation and comparison with video see-through technology. IEEE Access. 2020;**8**:169024-169035

[19] Ferrari V, Cutolo F. Letter to the editor: Augmented reality–guided neurosurgery. Journal of Neurosurgery. 2016;**125**(1):235-237

[20] Zinsser T, Schmidt J, Niemann H. A refined ICP algorithm for robust 3-D correspondence estimation. In: Proceedings 2003 International Conference on Image Processing (Cat. No.03CH37429). Barcelona, Spain: IEEE; 2003. pp. II-695

[21] Manevarthe B, Kalpathi R. Geometric correction for projection on non-planar surfaces using point clouds. In: ICDSC'18: Proceedings of the 12th International Conference on Distributed Smart Cameras, September 2018. ACM; 2018. pp. 1-6

[22] Fischer J, Bartz D, Strasser W. Occlusion handling for medical augmented reality using a volumetric phantom model. In: Proceedings of the ACM Symposium on Virtual Reality Software and Technology (VRST'04). New York, NY, USA: ACM. pp. 174-177

[23] Rong W. Projection-Based Spatial Augmented Reality for Interactive Visual Guidance in Surgery. [PhD Thesis]. National University of Singapore; 2013

[24] Maiero J, Kruijff E, Hinkenjann A, Ghinea G. Focus-plus-context techniques for picoprojection-based interaction. IEEE Transactions on Multimedia. 2017;**19**(7):1521-1530

Chapter 4

The State of Augmented Reality in Aerospace Navigation and Engineering

Pratik Pradhan, Mohsen Rostami, Jafer Kamoonpuri and Joon Chung

Abstract

The concept of Augmented Reality (AR) has existed in the field of aerospace for several decades in the form of Head-Up Display (HUD) or Head-Worn Display (HWD). These displays enhance Human-Machine Interfaces and Interactions (HMI^2) and allow pilots to visualize the minimum required flight information while seeing the physical environment through a semi-transparent visor. Numerous research studies are still being conducted to improve pilot safety during challenging situations, especially during low visibility conditions and landing scenarios. Besides flight navigation, aerospace engineers are exploring many modern cloud-based AR systems to be used as remote and/or AI-powered assist tools for field operators, such as maintenance technicians, manufacturing operators, and Air Traffic Control Officers (ATCO). Thanks to the rapid advancement in computer vision and deep neural network architectures, modern AR technologies can also scan or reconstruct the 3D environment with high precision in real time. This feature typically utilizes the depth cameras onboard or independent from the AR devices, helping engineers rapidly identify problems during an inspection and implement the appropriate solutions. Some studies also suggest 3D printing of reconstructed models for additive manufacturing. This chapter covers several aspects and potentials of AR technology in the aerospace sector, including those already adopted by the companies and those currently under research.

Keywords: augmented reality (AR), modern technology, aerospace engineering, human-machine Interface (HMI), flight navigation

1. Introduction

Augmented Reality (AR) technology has had one of the most significant impacts in the aerospace sector. Caudell and Mizell [1] first coined the term "Augmented Reality" to explain an optical see-through head-mounted display that superimposed and anchored computer-generated graphics in an aircraft manufacturing plant. The technology would track the user's head pose and place a Computer Aided Design (CAD) or other relevant information in a simplified format augmented over the user's visual field of the real world, hence naming it "Augmented Reality." While the name only existed three decades ago, the concept of AR existed long before then. Both aircraft Head-Up Display (HUD) and Head-Worn Display (HWD) existed long before that. In this chapter, we attempt to discuss the evolution of these technologies slightly differently than several existing literatures [2, 3] but also provide information on how it is evolving particularly in the navigation, engineering, and design sectors.

2. Flight navigation

Today, flying an aircraft depends on three factors: the machine, the controls, and instruments, and the human operator [4, 5]. The machine is what flies, and the human operator (i.e., pilot) is the one that flies the machine. But, without the proper controls and instrumentation, pilots would have no clue how or which direction to fly the aircraft in. Of course, the first powered aircraft by the Wright Brothers, the 1903 Wright Flyer that flew in Kitty Hawk, North Carolina, did not have any instruments to guide the pilots of such information [6]. Instead, the person on the ground would use a stopwatch, an anemometer, and an engine revolutions counter to calculate distance flown, speed, and horsepower of the propeller engine. Following the Wright Brother's invention, many would continue to fly the aircraft. However, without proper instrumentation, many pilots would lose their lives because of structural failure or stalling, leading to the implementation of the first visual indicators in 1907. Pilots would be trained to fly the Wright aircraft using an incidence indicator consisting of two limiting red marks on the scale to identify the relative pitch of the aircraft [5]. Mechanical displays would continue to evolve for the next 5–6 decades followed by the electromechanical displays between 1930s and 1970s, and then by the first and second generations of Electronic Flight Instrument System (EFIS) [7]. Mechanical displays were pressure-based instruments and would often result in slower than required indication of various flight parameters. Unlike them, the electromechanical instruments would be electrically powered while the indications would still be driven pneumatically [8]. After decades of research within civil and military aviation, a standard arrangement of various instruments was developed which is still used in old aircraft today. While such displays provided more stable and accurate data for pilots during flight, the need to have more information for better situational awareness would lead to the need to requiring more eyes on the flight deck, resulting in the development and evolution of the EFIS. EFIS is a purely digital display system that receives its data through the onboard flight computer which receives its data from the onboard sensors. The newer generation of EFIS, referred to as the glass cockpit, uses a standard set of display units including a Primary Flight Display (PFD), a Navigational Display (ND), an Engine Indicating and Crew Alerting System (EICAS), or an Electronic Centralized Aircraft Monitor (ECAM), a Multifunctional Display (MFD), and a Flight Management Computer (FMC) [9, 10].

Much like the evolution of HDDs, the first recorded usage of a HUD dates to the 1920s, used as a reflector gunsight in a fighter aircraft by Sir Howard Grubb. His design was important because the gunsight would project a distant virtual image of a back illuminated aiming graticule such that the graticule could be superimposed over the distant target. For a typical gunsight back then, the gunman would have to align the target with a backsight and a foresight. That said, it was not until the 1940s that a dynamic visual component would be added to a reflector gunsight. Maurice Hancock designed this gyroscopic gunsight and used it on the RAF Spitfire and Hurricanes. For his invention, he used two independent sights: one was a version of Grubb's sight, and the second was an aiming symbol that shifted across the line of sight by an angle that changed based on aircraft speed, altitude, attitude, and turn rate [9, 11]. Following this important feat, military aircraft in the 1950s and 1960s would begin displaying other flight-related details such as flight path vector into the displays. In 1962, a British strike aircraft named the Blackburn Buccaneer would be the first aircraft to have a fully operational HUD [2, 12]. By the 1970s, HUDs would start being used in commercial aircraft, starting with Sextant Avionique in the Dassault Mercure aircraft in 1975, shortly followed by Sundstrand and Douglas in their MD80 series aircraft. Once the technology hit the commercial market, HUDs were prioritized for safe landing and low visibility operations. By the early 2000s, HUD-equipped commercial aircraft had logged over 6 million flight hours with 30 thousand low visibility operations [13]. In 2009, the Flight Safety Foundation (FSF) released a report stating that the Head-Up Guidance System Technology (HGST) prevented about 38 percent overall potential accidents and 69 percent overall accidents caused during take-off or landing [13, 14]. Today, almost all the airlines and business jet aircraft are equipped with an HUD system. The evolution of HUD and VR would later inspire the invention of the Sword of Damocles by Ivan Edward Sutherland in 1968 [15], and the development of the Visually Coupled Airborne Systems Simulator (VCASS) [16] and the Super Cockpit Program [17, 18], both led by Thomas A. Furness III between 1960s and 1980s. Their work would inspire the military to consider the usage of Helmet Mounted Displays (HMDs) to be able to always visualize minimum flight and combat information during the flight.

2.1 Head-up display (HUD)

A HUD is comprised of two components: a Pilot Display Unit (PDU), and a HUD computer [13]. The PDU is simply a semi-transparent visor that is situated in the glareshield or above the pilot's head. The HUD computer generates an image based on the flight information which is then reflected onto the PDU through a projector connected to the computer. To ensure visibility throughout the various stages of flight, the displayed contents are usually either monochrome green or a combination of monochrome green and magenta. The combiner glass on the PDU is specially coated so that only the color of light projected from the image source is visible to the pilot.

The main purpose of a HUD is to superimpose imagery over the pilot's forward Field of View (FOV) outside the window [19]. In doing so, it reduces the amount of time pilots would have to focus on the HDD, especially during landing or low visibility conditions. HUD contents are being collimated on the visor which means the light rays are traveling parallel with the eye resulting in an infinite visual. Hence, the focus of the eyes would not need to be readjusted when transitioning between the display and the OTW. Lastly, the HUD's graphical contents are generated digitally. Hence, some modified components of the imagery can be conformed with what the visuals are trying to represent. For instance, in a taxiway, HUD can be adjusted to overlay conformed representation of the horizon line as seen on the OTW from the pilot's point of view as seen in **Figure 1**. Or it could be used to display advanced symbologies such as the Tunnel-in-the-Sky (TS) visual as later shown for a conceptual Urban Air Mobility (UAM) simulation in **Figure 2**.

Figure 1.
C-130 j HUD [20].

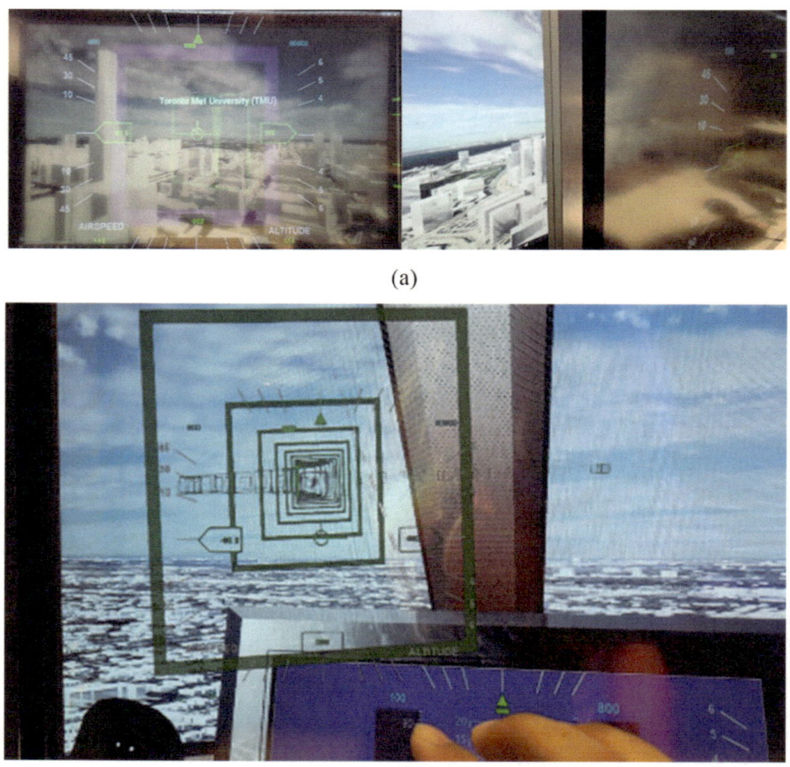

Figure 2.
The next generation of UAM AR-based cockpit (a) HUD view for UAM pilot's point of view on a transparent AR screen (b) HUD view for using Microsoft HoloLens 2 to proof the concept of the UAM flight corridors [21].

2.2 Helmet mounted Display (HMD)

When using the HUD, it is assumed that the pilot only needs to focus on his/her forward FOV. As shown in **Figure 3**, HUD's total FOV is much smaller than that of the HMD's. This is mainly because the HUD's total FOV is often the same as its instantaneous FOV as the pilot is assumed to be focusing only on the HUD. On the contrary, HMDs are equipped with the head tracking feature allowing the pilots to move around. Hence, their total FOV is much larger than their instantaneous FOV [22].

Although HMDs tend to provide better SA around the aircraft during flights, they are often prone to pilot discomforts. Imagine a pilot flying an aircraft with a HUD while only looking in one direction. Now, imagine the same pilot flying the same aircraft with an HMD while trying to look in the same direction. Since the HMD is locked to the pilot's head directly, his/her head also needs to be rigid, which is a difficult task for any living being. As a result, HMDs (similar to the Thales TopOwl HMD shown in **Figure 4**) are equipped only in military aircraft and not on any commercial aircraft.

That said, many aerospace officials have begun to rely on modern AR Head-Worn Displays (HWDs) for research and training purposes. The new generation of AR and XR headsets such as Varjo XR-3, Microsoft HoloLens, Magic Leap 2, etc. is not only capable of generating extremely high-resolution visuals but also capable of generating spatially anchored data in the close-proximity environment (**Figure 5**). While these devices are not certified for in-field navigation purposes, these have proven to be a great tool for pilot training [25], simulation [26], and HMI testing purposes (**Figure 6**) [4].

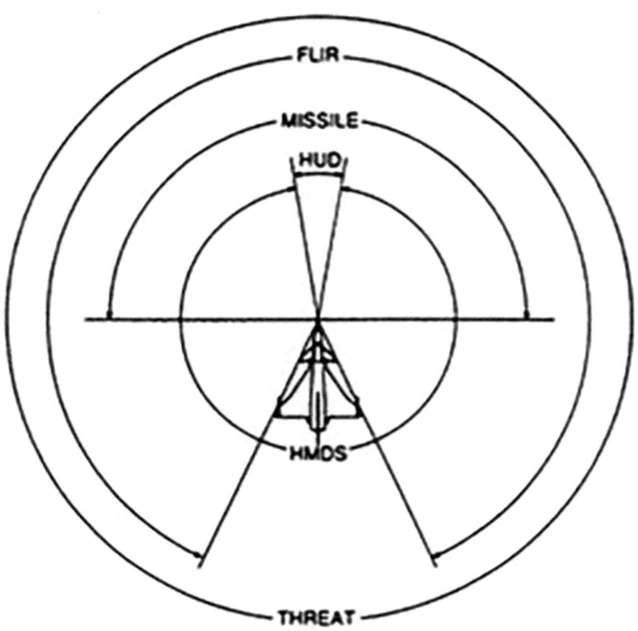

Figure 3.
HUD versus HMD: FOV [22].

Figure 4.
Thales TopOwl HMD [23, 24].

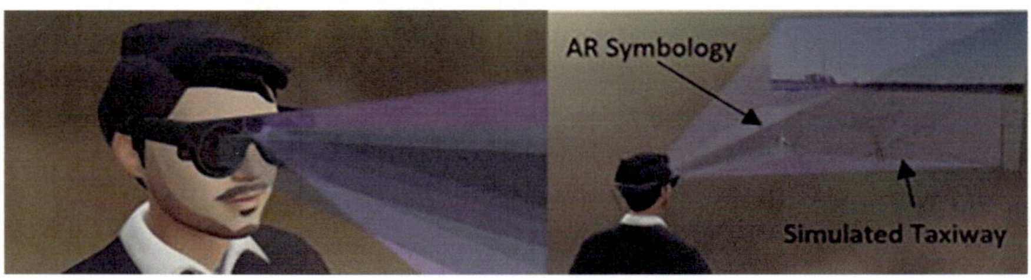

Figure 5.
Magic leap 2 AR headset being used for a potential application of airport surface navigation.

2.3 Degraded visual environment (DVE)

In aviation, one of the most dominant factors of aircraft accident is the Degraded Visual Environment (DVE), similar to the one shown in **Figure** 7 [28]. A degraded visual condition is referred to as a state in which pilots experience partial or complete loss of visual cues, often due to fog, time of day, brownouts, whiteouts, or simply due to bad weather [29]. Flights during such situations often result in reduced Situational Awareness (SA). As will be discussed in the next couple of sub-sections, pilots heavily rely on visual cues to taxi, take-off, and land an aircraft or a rotorcraft. However, if they cannot see these cues, they need to rely on the instruments. These rules are categorized as Visual Flight Rules (VFR) and Instrument Flight Rules (IFR). One of the key problems with IFR during DVE conditions is that pilots can experience spatial disorientation between the Out-The-Window (OTW) visuals and what they see on the Head-Down Displays (HDD). Even for experienced pilots operating an aircraft or a rotorcraft via IFR can be challenging. More often than not, a small fault in the instrument can also lead to a disaster as described in [30]. One way to mitigate this problem is to utilize the AR technology to overlay the runway or taxiway information along with relevant terrain data to increase the pilot's awareness. Moreover, a combination of these symbologies along with a properly crafted SVS can help pilots operate in VFR conditions even in a DVE state [31, 32].

Figure 6.
Flight display testing using Microsoft HoloLens 1 [4].

Figure 7.
DVE caused during helicopter landing on desert [27].

2.4 Vision systems

As mentioned in the previous section, in a DVE condition, while protocol dictates pilots to follow IFR when operating an aerial vehicle in relation to the ground, pilots can occasionally experience spatial disorientation resulting in potential accidents. To prevent this, HUDs and HWDs are often equipped with various vision system technologies. In aviation, the three most common systems include: Enhanced Vision, Synthetic Vision, and Combined Vision [13].

2.4.1 Enhanced vision system (EVS)

EVS, or Enhanced Flight Vision System (EFVS), uses onboard sensors and light emitters to improve the visibility of the OTW environment. These sensors or emitters could include a Forward-Looking Infrared (FLIR) sensor, a millimeter wave radar scanner, a millimeter wave radiometer camera, or a set of Ultraviolet (UV) sensors. Besides the typical flight information, EFVS data are presented in the HUD or HMD via an analog or a digital video format as recorded from front of the aircraft with the visibility enhancements. Since EFVS is basically using standard physical equipment to improve the visibility, it is not supported for all environmental conditions and poses a limit to how helpful it could be when used with a HUD or an HWD [11, 12].

2.4.2 Synthetic vision system (SVS)

Unlike EFVS, SVS uses a 3D rendering tool to generate surrounding terrain models using databases based on the Global Navigation Satellite System (GNSS) data for position, heading, and elevation data. Since the data is generated separately, similar to developing scenes on 3D development platforms, any geolocated features such as airport markers, obstacles, or runway features can be conformed onto the virtual terrain architecture. Moreover, since the terrain model is generated based on available data, it can be used in all weather conditions [13, 33, 34].

2.4.3 Combined vision system (CVS)

As the name suggests, CVS combines the details captured from the real-world view in the EFVS and superimposes them onto the models generated for the SVS. It allows for a selective blending between the two technologies while providing real-time synthetic data, resulting in potentially better situational awareness than either of the previous systems [22, 35].

2.5 Surface navigation

One of the most challenging aspects of aircraft navigation is taxiing it along the airport taxiway [36]. Especially for large aircraft, pilots must be able to ride the aircraft while following the taxiway centerlines precisely. Traditionally, pilots rely on verbal communication with the Air Traffic Control Officers (ATCOs) and taxi charts. Airport taxiways and runways are often equipped with a collection of pavement markings and designation signs. Both ATCOs and taxi charts make references to these markings and signs allowing pilots to follow the taxiway and runway prior to take-off or after landing. Besides these two, most aircraft are also equipped with Electronic Moving Maps (EMM) or Onboard Aircraft Navigation System (OANS) to assist pilots taxi more efficiently [37]. Despite some infrastructure built to enhance their capability to taxi the aircraft, a single miscommunication between the pilots and ATCOs, or their (pilots') misinterpretation of the taxi charts or maps can cause mild to fatal damage to the aircraft, its crew, and passengers, as reported in [38]. One way to minimize such incidents or accidents on the airport taxiways and runways is to use AR technology.

In 1996, David C. Foyle, et al. [39] introduced a HUD symbology configuration consisting of scene-linked 3D symbologies for taxiway centerlines and traffic edge cones and 2D symbologies for additional textual information such as Ground Speed (GS). These symbologies were designed to provide additional support to the pilots while minimizing their need to divert their attention to other visual contexts for the task and improve overall Situational Awareness (SA). Between 1996 and 2010, Foyle, Andre, and Hooey would lead multiple improvements on the design, focused on different aspects of the design such as importance of different types of information, or automated versus manual display of HUD components during a simulated flight.

One of the biggest challenges with surface flight operations using a HUD is that implementing head tracking is extremely complex as the scene-linked visual markers need to be relatively conformal to what they are representing. The simplest solution to this problem is to use a Head-Worn Display (HWD). Arthur et al. [40–42] led this area of research and implementation following the T-NASA study. Their concept for Beyond-RVR would allow them to view the scene-linked symbologies within a certain distance while still providing other flight information even if the pilots were to move their head around. An example of a similar concept is provided in **Figure 8**.

2.6 Air traffic control (ATC)

Potential avenues to enhance airport operations through the use of mixed reality have been proposed for decades, with a particular focus on air traffic control (ATC). This section will serve to highlight some of the noteworthy progress made in establishing a framework for mixed reality integration into ATC operations. In 2006, Reisman and Brown published a paper detailing the design of a prototype for augmented reality tools to be used in ATC towers. The Augmented Reality Tower Tool (ARTT) consisted of two phases; a prototype development and evaluation phase followed by an engineering prototype that resulted in the creation of a head-mounted display that superimposed simulated 3D images of runways, significant landmarks, and ATC data for the user to view and utilize to make decisions [43]. The system received mostly positive feedback from the ATC operators regarding its usefulness in a variety of tasks, including instances where coordination with aircraft under multiple low visibility scenarios was required. Another paper to note is Masotti's work on designing and developing a framework to prototype AR tools specifically for ATC tower operations [44]. Using augmented reality, Masotti proposed several benefits that included a reduction in the amount of visual scanning required and an increase in situation awareness due to the relevant information being superimposed on the real-world view for the ATC operators in an organized manner. AR tools implemented

Figure 8.
T-NASA display on a HUD [36].

in low visibility conditions aided in increasing situation awareness for operators and allowed for less time to be spent analyzing head-down operations.

Safi and Chung [3] provide a detailed exploration of AR applications and their uses in aerospace and aviation. Their chapter discusses the benefits and drawbacks of integrating AR into ATC operations. They also highlight the contrast between head-up and head-mounted displays (HUD and HMD respectively). While HUDs provide a larger field of view (FOV) and reduce computer processing lag through their direct connection to the terminals in the ATC tower, the lack of motion tracking capability and information only being accessible on these see-through displays limits the freedom of the ATC operator to move around and reduces their immersion. By contrast, HMDs solve these drawbacks but suffer with their limited FOV and discomfort during extended periods of wearing the HMD due to its weight. Moruzzi et al. have proposed the design and implementation of eye tracking application on a see-through display [45], with the goal being to achieve the concept of a Remote and Virtual Control Tower (RVT). Depending on the movement of each eye, digital content would then be overlaid onto the display where it would be the most appropriate and convenient for the user to view. Using a Microsoft Kinect device, the location of the eye on a human face was able to be distinguished and an algorithm tracked the motion of each user's eye to then determine the placement of digital information to be superimposed on the screen.

2.7 Urban air mobility (UAM)

Recent developments, particularly regarding electric propulsion and battery storage, have led to flying vehicle concepts for personal usage [46–48]. Urban Air Mobility (UAM) is the new air transport system that uses low-mid level urban airspace below ~2000 ft. UAM is a subset of the Advanced Air Mobility (AAM) under development by NASA, FAA, and industry [16, 17]. UAM focuses on the urban and suburban environments [49].

Currently, the main challenges UAM is facing are community acceptance, safety concerns, airspace management, and required advances in ATC and autonomy. And many companies such as Airbus/Boeing/Honeywell even organizations like NASA, EASA, and ICAO are involved to speed up the development and acceptance of UAM in modern countries. Considering the growing interest in AR technology, rapid growth of UAM industry requires the incorporation of AR. Accordingly, instead of dealing with the physical controls, everything will be digital and imaginary, and flight mechanics and dynamics of the aircraft will be shown on the AR screen using the concepts of Human-Machine Interfaces and Interactions (HMI2) through symbology design while projecting the orientation and position of the UAM aircraft. Furthermore, using this new technology, the required time to train UAM pilots will significantly drop compared to the existing technologies, as the integrated AR system could be used from start to finish to train the pilots.

UAM needs to be easily, safely, and semi-automatically operated to be accepted by the public. Accordingly, three major areas could be targeted including flight monitoring, operation, and training. Among these, particularly monitoring and operation of UAM aircraft could be enhanced by AR which will be discussed in the following.

The monitoring is mostly about the next generation of control tower concepts that would benefit from AR [45, 50, 51]. For the purpose, traffic visualization, predictable corridors, and automated mission management are the key aspects of semi-autonomous operation of UAM aircraft. All information related to the flight, aircraft

DOI: http://dx.doi.org/10.5772/intechopen.1002358

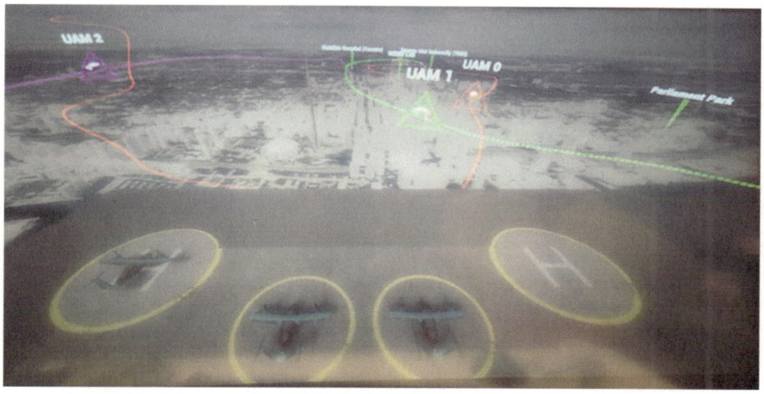

Figure 9.
View from the next generation of UAM control tower concepts that will be using AR windows [57].

specifications, weather data, and safety of the airspace and each air vehicle should be shown in real time on the AR windows and presented to the tower operator [52–54]. This will also be part of the tasks that are defined for the Providers of Services for UAM (PSU) that are responsible for the operations planning, flight intent sharing, airspace management functions, off-nominal operations, operations optimization, and airspace reservations [55, 56]. **Figure 9** presents the new AR-based ATC system using a transparent AR screen for UAM flight monitor in an urban environment [57].

Advances in Airspace Management and Automation for UAM operation are must due to safety concerns. The operation of UAM aircraft can also benefit from AR-based cockpits to modernize flight corridors according to the safety concerns and risk evaluation methods, augment the flight by auto-generation of new routes, and approve the new route before execution, use optimized flight path where autonomously adapt flight considering weather conditions and other aircraft, train, and obstacles (**Figure 2**) [21].

3. Engineering operations

AR-based instructions are a series of visual information to guide users through an assembly process for complex engineering products [58, 59]. Due to the technological enhancements and the highly competitive engineering environment, innovative products are required to be taken into the market in a short period of time. This also demands a collaborative manufacturing environment to exchange real-time information [60]. Nowadays, thanks to the efforts of Airbus, Boeing, Bombardier, Safran, and Siemens, the new technology has been successfully applied in design, manufacturing, assembly, Maintenance, Repair, and Overhaul (MRO), and education. In this section, AR usage for design and assembly, collaborative interface, Artificial Intelligence (AI) and Machine Learning (ML) implementations, haptic integration, and computer vision implementations are discussed (**Figure 10**).

3.1 Design and assembly

Descriptive assembly instructions for complex procedures are a key factor to ensure a smooth assembly [62]. AR presents information in the cyber-physical space

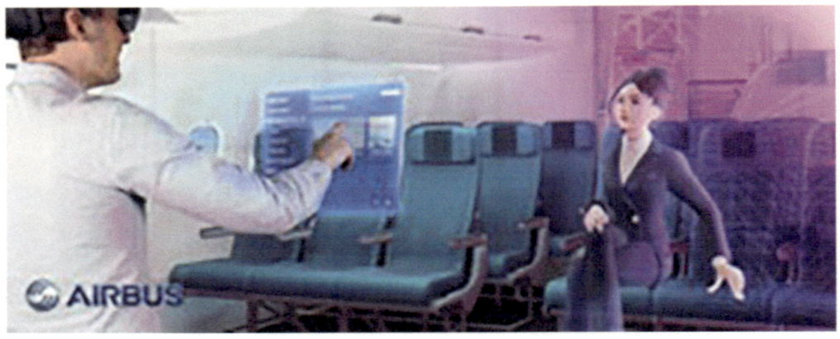

Figure 10.
Airbus uses immersive collaboration concept for cabin definition [61].

in the form of virtual models [63] and interacts with the actual assembly parts [64]. The application of assembly instructions has changed from just providing assembly procedures to actively providing heuristic visual guidance to meet operators' cognitive requirements [65].

AR assembly has several benefits in enhancing the efficiency of manual assembly process. It mixes digital instructions and physical tasks. First, AR-based instructions extend the user's visual understanding from the actual world to the information space [66]. Engineers can communicate with actual objects in cyber-physical space using interactive tools [67]. This communication constantly improves the user's understanding of the actual model and deepens the usability and reliability of assembly procedures [68]. Secondly, compared to conventional assembly instructions, AR assembly instruction embeds interactive virtual instructions into the actual environment, enriching engineer's experience of the actual world [69]. Moreover, AR assembly instruction is economically efficient and straightforward to construct procedures. Data calculation is accomplished by the digital system and is only responsible for the construction and representation of the instruction materials [70]. Thirdly, the rule of engineer cognition exists which enhances the communication efficiency between engineer and system [71, 72]. Earlier assembly instruction materials include paper-based manuals and drawings, recorded videos, simulation animations, etc. **Figure 11** shows a graphical representation of the traditional assembly procedures in aerospace industries using physical manuals. Engineers usually experience a very time-consuming procedure, and the cognitive productivity is so low.

Nevertheless, the AR assembly instruction can enhance the interpretation of the engineer's operations perspective. Finally, compared to traditional manuals, AR-based assembly instructions have more natural, intuitive interactivity, and user performance [73]. Engineers can precisely collaborate with AR-based assembly instructions using bare-hand feedback [74]. These instructions are presented in a 2D graphics language, which combines two illustrations. The first depicts the real model which includes the part or assembly illustrated by its orthographic projection. The second illustration overlays instructions on the first using a standard-based language that defines dimensions, tolerances, and assembly details [75].

In 2003, to meet the requirements of 3D assembly instructions as a manufacturing source, the ASME Y14.41 standard was issued [76]. Hence, 2D engineering drawings are now permitted for engineering operations. Thanks to the development of digital assembly and design technology, information can be directly inserted into the CAD model to create a model-based definition (MBD) database [77], which includes 3D geometry and product dimensional tolerance, to achieve complete product illustration [78]. In 2001, for the first time, Boeing implemented enhanced digital manufacturing

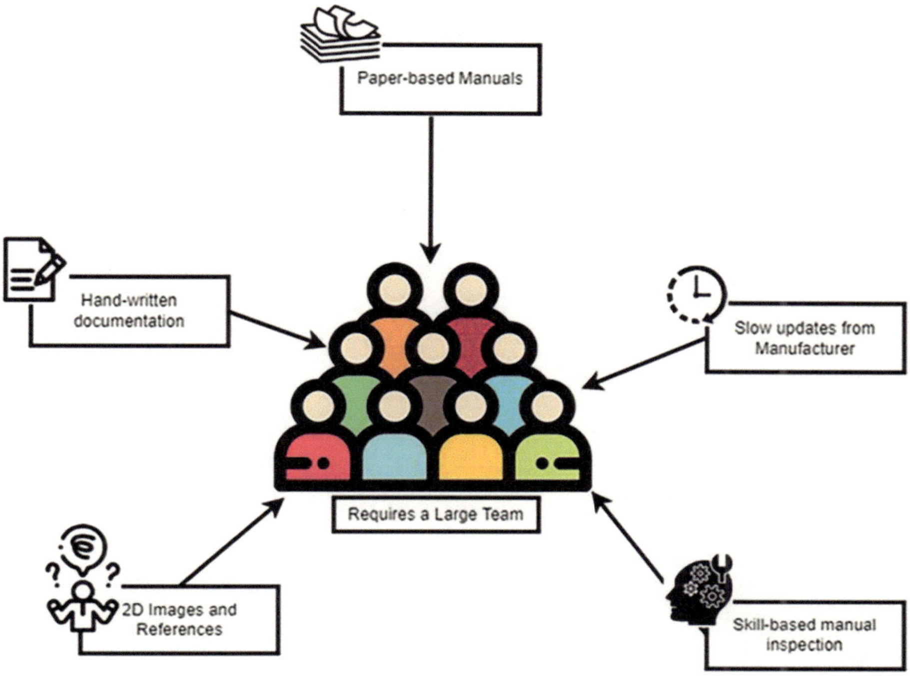

Figure 11.
Graphical representation of traditional assembly procedure in aerospace industries.

technology for the assembly process of aircraft cables [79]. Afterward, it applied digital commands to help the entire aircraft assembly line [80].

However, 3D AR assembly instructions still face some drawbacks. First, the spatial information reflected by 3D instructions could be confusing as the display of the blueprint is typically presented on a 2D display [81]. The engineer is required to continuously switch attention between the display and the actual task, which can distract the engineer and result in the failure of the task. Additionally, in the assembly process, the actual environment and the instructions in digital environment are totally separated [82]. Also, there are some challenges with the Host Controller Interface (HCI) mode of 3D assembly instructions. For instance, engineers regularly do not use their hands to directly manipulate the tasks but instead use 2D interfaces such as displays, mouse, and keyboard [83].

Figure 12 shows a graphical representation of the current assembly procedures in aerospace industries where tablet-based manuals are used along with cloud-based updates from manufacturers. Cloud-based assembly platform provides flexible, high-performance, and universal capabilities. It works with big data to support communication, share data, and answer engineer questions through information provided by the manufacturer and reliable sources [84, 85].

3.2 AI and ML implementation

Boeing 777 was the first aircraft that was designed entirely from simulation without a physical mock-up. Likewise, ML and AI algorithms are playing a great role in the future of Aerospace design and assembly [86–90]. According to the study

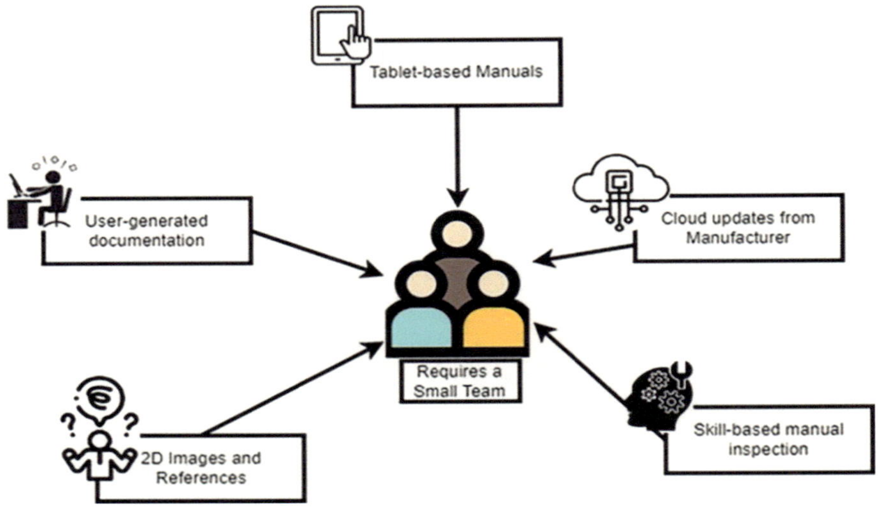

Figure 12.
Graphical representation of current assembly procedure in aerospace industries.

accomplished by Zhiwu AIoT Industry Research Institute, the Industrial Metaverse is a new ecosystem in which new information and communication technologies represented by the Internet of Things (IOT), AI, and digital twins are deeply integrated with the real economy [91].

ML is a growing set of optimization algorithms and regression methods to create models from data [86–92]. AI relies on the ability of computing to repeat human learning. It starts with mimicry by feeding large quantities of data to a network of neurons. It continues developing the network up until it can regenerate human abilities for reflection. In the next step, AI identifies responses in model data, images, synthesize information, predict trends, and present precise findings [93].

Each phase of advanced aerospace manufacturing is data-intensive, including design and assembly, testing, and service. A Boeing 787 contains millions of parts and subparts that are manufactured from around the world and assembled in an enormously complex manufacturing procedure. This results in massive multimodal information from supply chain logs, videos, inspection data, engineering drawings, and notes. After design and assembly, a single flight test will collect information from hundreds of thousands of multimodal sensors. In service as well, the aircraft creates numerous real-time data, which is collected, transferred, and processed with kilometers of wire and millions of code lines. Hence, big data is a reality in aerospace engineering and advanced data analytics with AI and ML is a must [87].

3.3 Embedded AR-assisted training

Haptic feedback would allow the engineer to directly perceive the data about the environment using the sense of touch [94, 95]. The term "Haptics" was proposed by Resvez in 1950 after observing a blind performance. The term refers to an unconventional sensory experience, deviating from traditional methods of touch and kinesthetics [96]. Haptic feedback is an alternative visual representation triggering condition. Haptic AR potentially overcomes the existing visual issues of AR allowing the user to focus on the task and avoid over-reliance on the technology [97].

Figure 13 graphically represents the concept of Haptic AI-assisted AR-based engineering operations in aerospace industries. The expected core technology instruction is as follows [98]:

1. Virtual-real interactive environment: The concept is supported by a semi-immersive virtual-real interactive space, which stimulates engineers' learning capacities. This allows aerospace engineers to feel all work types in the physical space and virtual world simultaneously, to allow full use of the surrounding real objects as communication/interaction elements to accomplish manual tasks.

2. Spatial relation rendering: To make AI-assisted AR-based engineering operations more realistic, the consistency between virtual objects and actual space should be clarified. This is about the consistency of the actual environment in geometric aspects such as position, perspective, and occlusion.

3. Physical cue rendering: Lighting consistency is a crucial element in ideal combination of AR-based environment and physical scenes. Based on the light distribution in the actual scene, shadow processing is completed on AR commands to compensate the floating sense of command material in vision. In a sense, human eyes have strong feedback clues. If the light does not match the actual scene, the engineer instantly feels that.

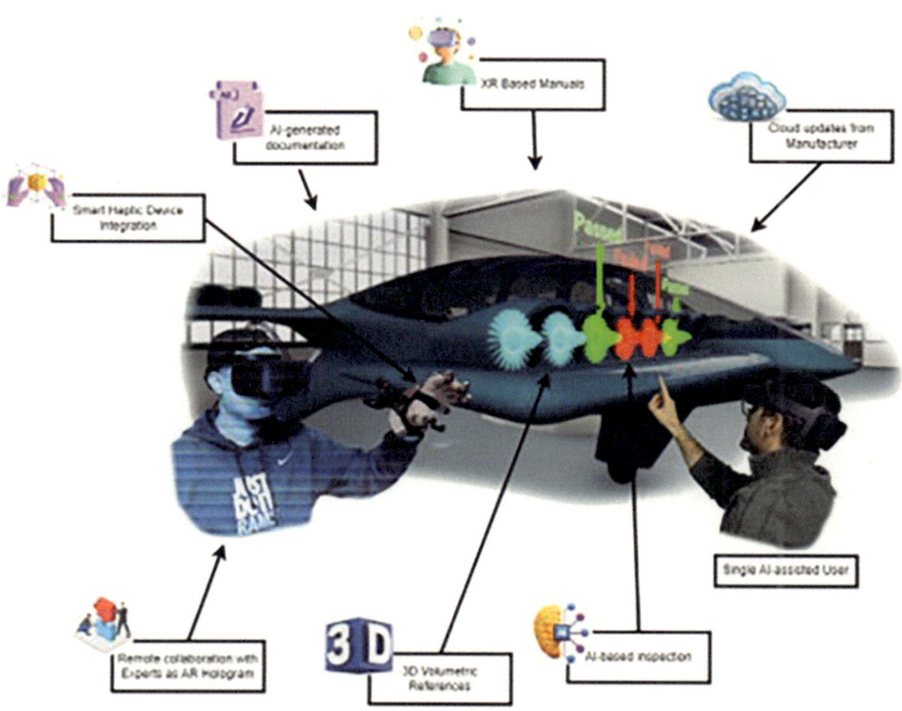

Figure 13.
Graphical representation of next generation of AR-based engineering operations in aerospace industries.

4. Cognitive cue rendering: To effectively accomplish the manual tasks, the instruction reflected in AR must be related to the engineer's cognition. The engineer must understand the intent of a physical process using the information provided in the AR-based instructions. In this process, assembly instructions are passed between the designer (or AI) and the engineer until an agreement is reached. Therefore, intention consistency is the most core feature of AI-assisted AR-based engineering operations.

3.4 Computer vision implementations

Computer vision is a field of artificial intelligence (AI) that trains a computer to extract information from images or video data. It involves the development of algorithms, techniques, and methodologies that enable computers to analyze, interpret, and understand visual data [99]. Computer vision algorithms support several use cases in aerospace engineering applications. For example, it can be used in quality control to inspect components, identify defects, and ensure adherence to quality specifications. Or, combined with AR technologies, can be used to develop training tools to provide real-time overlays of information, instructions, or virtual representations to guide the user through an assembly process.

One of the most common use cases of computer vision is in aircraft MRO for visual inspection. Using deep learning algorithms, technicians can scan the surface of an aircraft to identify potential maintenance issues. The convolutional neural network (CNN) provides the means to accomplish this. A CNN is a supervised deep learning model that is used for image classification. Supervised learning is a training strategy where a model is taught how to make predictions based on labeled examples. In this method, the model is provided with a set of input data, along with the correct answer or outcomes associated with that data. For image classification, the model extracts relevant features or patterns from images, which can include edges, textures, shapes, colors, and other visual characteristics. These are used to make predictions on new examples. There are various types of CNN architectures that exist, and it is the most common computer vision solution used in aerospace for a variety of applications [99].

A CNN is a supervised deep learning model that is used for image classification. Supervised learning is a training strategy where a model is taught how to make predictions based on labeled examples. In this method, the model is provided with a set of input data, along with the correct answer or outcomes associated with that data. For image classification, the model extracts relevant features or patterns from images, which can include edges, textures, shapes, colors, and other visual characteristics. These are used to make predictions on new examples. There are various types of CNN architectures that exist, and it is the most common computer vision solution used in aerospace for a variety of applications.

Identification of fuselage defects and visual checks are commonly addressed topics in the aircraft inspection literature. Autonomous visual inspection of an aircraft exterior is possible using drones and a high-resolution camera [100]. In 2021 German aerospace company, Lufthansa Technik trained a deep neural network to identify fuselage defects (i.e., dents, scratches) at the Lufthansa Technik summit. This was accomplished by capturing high-resolution photos of the aircraft exterior using a drone along path outlined in **Figure 14**. The snake-like pattern pans both sides and the nose of the plane, while the roof of the aircraft is captured using a downward-facing camera [102]. The captured images are sent to a computer for data processing and

Figure 14.
Simulated drone path [101].

identification of damages and irregularities [100]. Generating sufficient examples to adequately train a neural network was the biggest hurdle identified by Lufthansa, as there are no public datasets available for damage samples. Moreover, it is expensive and laborious to create their own training material. So, the neural network was trained on a synthetic dataset. The aircraft was simulated virtually where the hangar lighting, time of day, and placement of defects on the simulated aircraft are adjusted to generate a sufficient training dataset [102]. The resulting model yielded a detection accuracy above 95%, in the simulated environment with a dataset size of 4000 images.

A robust computer vision application depends largely on the quality and quantity of training data. Boeing used a computer vision approach to develop an AR application for aircraft inspection. The application would overlay markers atop the aircraft identifying points of existing damage marked by other mechanics. To accomplish this, the application needed to anchor a 3D virtual model on top of the aircraft as seen by the mechanic through the AR device. Boeing used a machine learning approach to solve this problem, generating their own dataset consisting of thousands of images of aircraft from different perspectives. Using a pose estimation algorithm, and a series of markers placed around the aircraft, the aircraft's position was calculated in relation to the camera. The machine learning model was fed this data to calculate the position of the aircraft based on the images coming from the camera without markers. Due to the small dataset size, the model performed poorly. And, like Lufthansa Technik, Boeing showed that using field data to train their machine learning model was ineffective due to the laborious nature of acquiring data [103].

Another example of innovative computer vision implementations in aerospace is the Airbus Wayfinder Project. The project was initiated by Airbus in June 2018 in collaboration with Project Wayfinder with the ambition to revolutionize autonomous flight [104]. The objective of this project was to achieve autonomous taxiing, take-off, and landing (ATTOL) of a commercial aircraft through a computer vision-based system. Although commercial aircraft can already fly an approach, land, and rollout to taxi without pilot input, this is only possible on runways equipped with a CAT III ILS. The installation and maintenance of this system is costly, and only a few airports can justify its cost. So, a deep learning model was used to teach an aircraft to fly an approach and landing sequence without any pilot input. The algorithm implements an object detection model to identify the runway, and a regression model to estimate the distance, localizer, and glide slope values of the aircraft [105]. A modified Single Shot

Detection network (SSD) was used to accomplish this. The beginning layers of the SSD are used to extract features about the runway environment, while the ending layers are responsible for boxing the region containing the runway, distance to the runway, localizer deviation, and glideslope deviation. Like others, Airbus relied on synthetic data and developed a model that can reliably detect the runway from several miles away.

4. 3D reconstruction and visualization

3D reconstruction is the process of capturing the shape and appearance of real objects. 3D reconstruction can be accomplished by numerous methods. Single cameras are computationally efficient but require other sensors to determine depth scale. Stereo cameras utilize images captured from two cameras set a defined distance. An algorithm is used to evaluate the depth using the two images. Lastly, RGBD cameras, such as LiDAR, can perform range detection using structured light sensors to directly capture depth information [106].

In recent years, LiDAR technology has shown more applications with its capabilities for remote sensing and data acquisition. LiDAR technology provides a unique advantage over traditional remote sensing methods through high-resolution data acquisition with spatial and real-time capabilities. LiDAR sensors can construct detailed point clouds representing the shape, structure, and surface characteristics of objects and landscapes.

LiDAR technology has shown promising results for the inspection of airframes and aerodynamic surfaces [107]. The ability to capture highly detailed 3D representations of aircraft components and structures facilitates advanced inspection techniques. The traditional method of visual inspection, although common, has limitations in detecting these defects due to the lack of contrast and reflectance on most surfaces. Moreover, the reliance on human inspections and specialized equipment increases inspection time and cost significantly. By obtaining a 3D point cloud of the aircraft's parts and comparing it with a reference CAD model, surface deformations can be detected. This comparison generates a disparity map that highlights the differences between the reference CAD model and the inspection point cloud. With current LiDAR technologies, reconstruction accuracies can be obtained with errors less than 1 millimeter.

3D reconstruction of the environment is another important use case in aerospace, primarily for analyzing terrain. With the advance of drone technology, practical applications of 3D reconstruction include the inspection and mapping of areas that are difficult to access by humans. For instance, the mapping of an accident site can be accomplished using a high-resolution camera and 3D reconstruction tools [108]. Using LiDAR technology, this is achieved by relating the RGB frame with its depth frame. The simultaneous localization and mapping algorithm, also known as SLAM, is one such method that builds a map of the environment while localizing the camera in the map at the same time [109]. SLAM allows aircraft to map out unknown environments, which is extremely useful to carry out tasks such as path planning and obstacle avoidance.

5. Conclusion

In conclusion, this chapter presented the current applications of AR in the field of aerospace, particularly in navigation and engineering. AR has been used in aerospace navigation for several decades in the form of HUD or HWD. These displays enhance HMI^2 and allow pilots to navigate the aircraft flight. Besides navigation, aerospace engineers are exploring AR systems to be used as remote and/or AI-powered assist tools for field operators, such as MRO, design, and assembly and ATCO. AR technologies can also scan or reconstruct the 3D environment with high resolution in real time using computer vision and deep neural network architectures to allow engineers to rapidly identify problems during an inspection.

Acknowledgements

We would like to thank Niraj Karki and Charith Kongara for providing their knowledge and insights in ATC and Computer Vision, and thoroughly reviewing the drafts along the way.

Appendices and nomenclature

AAM	advanced air mobility
AI	artificial intelligence
AR	augmented reality
ARTT	augmented reality tower tool
ATC	air traffic control
ATCO	air traffic control officer
ATTOL	autonomous taxiing, take-off, and landing
CAD	computer aided design
CNN	convolutional neural network
CVS	combined vision system
DVE	degraded visual environment
EASA	european union aviation safety agency
ECAM	electronic centralized aircraft monitor
EFIS	electronic flight instrument system
EFVS	enhanced flight vision system
EICAS	engine indicating and crew alerting system
EMM	electronic moving map
EVS	enhanced vision system
FAA	federal aviation administration
FMC	flight management computer
FOV	field of view
FSF	flight safety foundation
GS	ground speed
HDD	head-down display
HGST	head-up guidance system technology
HMD	helmet-mounted display
HMI	human-machine interface
HMI2	human-machine interface and interaction
HUD	head-up display
HWD	head-worn display
ICAO	international civil aviation organization
IFR	instrument flight rules

IoT	internet of things
LiDAR	light detection and ranging
MFD	multifunctional display
ML	machine learning
MRO	maintenance, repair, and overhaul
NASA	national aeronautics and space administration
OTW	out-the-window
PDU	pilot display unit
PFD	primary flight display
PSU	provider of services for UAM
RGBD	red green blue - depth
RVT	remote and virtual tower
SA	situational awareness
SLAM	simultaneous localization and mapping
SSD	single shot detection
SVS	synthetic vision system
T-NASA	taxiway navigation and situational awareness
TS	tunnel-in-the-sky
UAM	urban air mobility
VCASS	visually couple airborne systems simulator
VFR	visual flight rules
VR	virtual reality
XR	extended reality

Author details

Pratik Pradhan*, Mohsen Rostami, Jafer Kamoonpuri and Joon Chung
Toronto Metropolitan University, Toronto, Canada

*Address all correspondence to: pratik.pradhan@torontomu.ca

References

[1] Caudell TP, Mizell DW. Augmented reality: An application of heads-up display technology to manual manufacturing processes. In: Proceedings of the Twenty-Fifth Hawaii International Conference on System Sciences. Vol. 2. Kauai, HI, USA: IEEE; 1992. pp. 659-669. DOI: 10.1109/HICSS.1992.183317

[2] Safi M, Chung J, Pradhan P. Review of augmented reality in aerospace industry. Aircraft Engineering and Aerospace Technology. 2019;**91**:1187-1194. DOI: 10.1108/AEAT-09-2018-0241

[3] Safi M, Chung J. Augmented reality uses and applications in aerospace and aviation. In: AYC N, Ong SK, editors. Springer Handb. Augment. Real. Cham: Springer International Publishing; 2023. pp. 473-494. DOI: 10.1007/978-3-030-67822-7_20

[4] Pradhan P. Augmented Reality Cockpit Display System in Real-Time Flight Simulation Environment [thesis]. Toronto, ON, Canada: Toronto Metropolitan University; 2023 [cited 2023 Jul 6]. DOI: 10.32920/23580471.v1

[5] Coombs LFE. Control in the Sky: The Evolution & History of the Aircraft Cockpit. Barnsley, SY, UK: Pen & Sword; 2005

[6] Wright Flyer. 1903 National Air and Space Museum n.d. Available from https://airandspace.si.edu/collection-objects/1903-wright-flyer/nasm_A19610048000 [accessed July 6, 2023]

[7] Lim Y, Gardi A, Sabatini R, Ramasamy S, Kistan T, Ezer N, et al. Avionics human-machine interfaces and interactions for manned and unmanned aircraft. Progress in Aerospace Science. 2018;**102**:1-46. DOI: 10.1016/j.paerosci.2018.05.002

[8] Nicholl R. Airline Head-up Display Systems: Human Factors Considerations. Saarbrücken: LAP LAMBERT Academic Publishing; 2015

[9] Federal Aviation Association. FAA-H-8083-15B, Instrument Flying Handbook. Oklahoma City, OK, USA: United States Department of Transportation; 2012. Available from: https://www.faa.gov/sites/faa.gov/files/regulations_policies/handbooks_manuals/aviation/FAA-H-8083-15B.pdf

[10] Jarrett DN. Cockpit Engineering. Aldershot, Hampshire, England. Burlington, VT: Ashgate; 2005

[11] Aukstakalnis S. Practical Augmented Reality: A Guide to the Technologies, Applications, and Human Factors for AR and VR. Boston: Addison-Wesley Professional; 2017

[12] Nijboer D. Fighting Cockpits: In the Pilot's Seat of Great Military Aircraft from World War I to Today. Minneapolis, MN, USA: Zenith Press; 2016 [cited 2023 Jul 6]. Available from: http://archive.org/details/fightingcockpits0000nijb

[13] Wood RB, Howells PJ. Head-Up Display. In: Spitzer CR, Ferrell U, Ferrell T, editors. Digital Avionics Handbook. 3rd ed. Bowie, MD, USA: CRC Press, Taylor & Francis Group; 2015. pp. 17-1-17-27

[14] Head-Up Guidance System Technology — A Clear Path to Increasing Flight Safety n.d.:29

[15] Sutherland IE. A head-mounted three dimensional display. In: Proc. Dec. 9-11 1968 Fall Jt. Comput. Conf. Part - AFIPS 68 Fall Part I. San Francisco, California: ACM Press; 1968. p. 757. DOI: 10.1145/1476589.1476686

[16] A visually-coupled airborne systems simulator (VCASS) - An approach to visual simulation. n.d.

[17] Livingston MA, Rosenblum LJ, Brown DG, Schmidt GS, Julier SJ, Baillot Y, et al. Military applications of augmented reality. In: Furht B, editor. Handb. Augment. Real. New York, NY: Springer New York; 2011. pp. 671-706. DOI: 10.1007/978-1-4614-0064-6_31

[18] Furness TA. The super cockpit and its human factors challenges. Proc Hum Factors Soc Annu Meet. 1986;**30**:48-52. DOI: 10.1177/154193128603000112

[19] Wickens CD, Ververs PM, Fadden S. Head-up displays. In: Harris D, editor. Human Factors for Civil Flight Deck Design. 1st Ed. Oxfordshire, UK: Routledge; 2004. DOI: 10.4324/9781315253039

[20] Logistics T. English: C-130J: Co-pilot's Head-Up Display. California: Former Castle AFB; 2005

[21] Rostami M. Development of an Extended Reality Based Flight Simulator for Urban Air Mobility. Conference presentation presented at: SETP Canadian Section Symposium; Ottawa, ON, Canada; May 2023

[22] Collinson RPG. Displays and man–machine interaction. In: RPG C, editor. Introd. Avion. Syst. Dordrecht: Springer Netherlands; 2011. pp. 19-99. DOI: 10.1007/978-94-007-0708-5_2

[23] Tiraden Français. Viseur de casque Topowl de la société Thales. commandé par plusieurs pays dont l'armée de terre française pour les pilote de l'Eurocopter Tiger. Photo prise au salon du Bourget 2017. 2017.

[24] TopOwl. Helmet-mounted Sight & Display for Helicopters. Thales Group. n.d. Available from: https://www.thalesgroup.com/en/markets/aerospace/flight-deck-avionics-equipment-functions/helmet-mounted-display/TopOwl [accessed July 6, 2023]

[25] Arjoni DH, de Souza RI, Pereira Figueira JM, Villani E. Augmented reality for training formation flights: An analysis of human factors. Heliyon. 2023;**9**:e14181. DOI: 10.1016/j.heliyon.2023.e14181

[26] Varjo. Virtual and Mixed Reality for Pilot Training and Simulation. Washington, DC, USA: Varjo; [cited 2023 Jul 6]. Available from: https://25667574.fs1.hubspotusercontent-eu1.net/hubfs/25667574/eBook%20and%20whitepaper%20PDFs/Varjo_Whitepaper_PilotTraining%20(1).pdf?utm_campaign=Content%20Downloads&utm_medium=email&_hsmi=63184198&_hsenc=p2ANqtz-92EGVID4937b3iqx8VWwKm2S360mbOHG-DXyjav3qMiJSI2xHkDpVEa--sNsy794JTC0Jx9WdgbqU5ogyMEg7T9dLeBdBcd525cVZCrdrbscZAaNw&utm_content=63184198&utm_source=hs_automation

[27] KUWAIT – Lt. Caleb Wyman and Lt. Cmdr. Def Vis Inf Distrib Serv 2015. Aug. 19, 2015 Available from: https://nara.getarchive.net/media/kuwait-aug-19-2015-lt-caleb-wyman-and-lt-cmdr-aa76af [accessed July 6, 2023]

[28] Szoboszlay Z, Miller J, Godfroy-Cooper M, Davis B, Feltman K, Hartnett RG, et al. The Design of Pilot Cueing for the Degraded Visual Environment Mitigation (DVE-M) System for Rotorcraft. In: Crew Stations and Human Factors [Online]. The Vertical Flight Society; 2021 [cited 2023 Jul 6]. Available from: https://vtol.org/store/product/the-design-of-pilot-cueing-for-the-degraded-visual-environment-mitigation-dvem-system-for-rotorcraft-16746.cfm

[29] Chittaluri V. Development and Evaluation of Cueing Symbology for Rotorcraft Operations in Degraded Visual Environment (DVE). [thesis]. Toronto, ON, Canada: Toronto Metropolitan University; 2022 [cited 2023 Jul 6]. DOI: 10.32920/ryerson.14665293.v1

[30] Spatial Disorientation Accidents: IFR in IMC 2020. https://www.aopa.org/training-and-safety/online-learning/safety-spotlights/spatial-disorientation/spatial-disorientation-accidents-ifr-in-imc [accessed July 6, 2023]

[31] Mubarak SNHF, Jacob AA. The impacts of advanced Avionics on degraded visual environments. International Journal of Aviation, Aeronautics, and Aerospace. 1 Jan 2023;**10**(1):1-12 [cited 2023 Jul 6]. DOI: 10.58940/2374-6793.1773

[32] Behringer R, Tam C, McGee J, Sundareswaran V, Vassiliou M. A System for Synthetic Vision and Augmented Reality in Future Flight Decks. In: Proceedings of the SPIE. Orlando, FL, USA: SPIE Aerospace; 2000. p. 81-86. DOI: 10.1117/12.389332

[33] Shelton KJ, Kramer LJ, Ellis K, Rehfeld DSA. Synthetic and enhanced vision systems for nextGen (SEVS) simulation and flight test performance evaluation. n.d.

[34] Arthur JJ III, Prinzel LJ III, Kramer LJ, Parrish RV. Flight Simulator Evaluation of Synthetic Vision Display Concepts to Prevent Controlled Flight into Terrain (CFIT). Hampton, Virginia: NASA: Langley Research Center; 2004

[35] Kratchounova D, Newton D, United States. Department of Transportation. Federal Aviation Administration. Office of Aviation. Civil Aerospace Medical Institute. Combined Vision Systems Literature Review. 2019.

[36] Hooey B, Foyle D, Andre A. A Human-centered methodology for the design, evaluation, and integration of cockpit displays. 2002

[37] Guilloton A, Arethens J-P, Avionics T, Macabiau C, Koenig D. State of the art in airport navigation. In: 2011 IEEEAIAA 30th Digit. Avion. Syst. Conf., Seattle, WA, USA: IEEE; 2011. pp. 4B3-1-4B3-11. DOI: 10.1109/DASC.2011.6096072

[38] NTSB Search Results [Internet]. Washington, DC, USA: FAA Aviation Safety Information Analysis and Sharing (ASIAS); [cited 2023 Jul 5]. Available from: https://www.asias.faa.gov/apex/f?p=100:27:::NO:27

[39] Foyle DC, Andre AD, McCann RS, Wenzel EM, Begault DR, Battiste V. Taxiway navigation and situation awareness (T-NASA) system: Problem, design philosophy, and description of an integrated Display suite for low-visibility airport surface operations. SAE Transactions. 1996;**105**:1411-1418

[40] Bailey RE, Shelton KJ, Arthur III JJ. Head-worn displays for NextGen. In: Marasco PL, Havig PR, editors. Head- and Helmet-Mounted Displays XVI: Design and Applications. Orlando, FL, United States: SPIE Digital Library; 2011. p. 80410G [cited 2023 Jul 5]. DOI: 10.1117/12.885847

[41] Trey, Jarvis AJ, Bailey RE, Williams SP, Prinzel LJ, Shelton KJ, Jones DR, et al. Review of head-worn displays for the next generation air transportation system. Optical Engineering. 2017;**56**:051405-051405. DOI: 10.1117/1.OE.56.5.051405

[42] Arthur JJ, Bailey RE, III Prinzel LJ, Kramer LJ, Williams SP. Multi-modal cockpit interface for improved airport surface operations. US7737867B2. 2010

[43] Reisman R, Brown D. Design of Augmented Reality Tools for air traffic control towers. In: 6th AIAA Aviat. Technol. Integr. Oper. Conf. American Institute of Aeronautics and Astronautics: ATIO, Wichita, Kansas; 2006. DOI: 10.2514/6.2006-7713

[44] Masotti N, De Crescenzio F, Bagassi S. Augmented reality in the control tower: A rendering pipeline for multiple head-tracked head-up displays. In: De Paolis LT, Mongelli A, editors. Augment. Real. Virtual Real. Comput. Graph. Cham: Springer International Publishing; 2016, p. 321-338. DOI: 10.1007/978-3-319-40621-3_23.

[45] Moruzzi MC, Santhosh S, Corsi M, Bagassi S, De Crescenzio F. Design and implementation of eye tracking application for generation of augmented reality content on spatial see through display of remote and virtual control tower (RVT). International Journal on Interactive Design and Manufacturing (IJIDeM). 2023;**17**:1859-1867. DOI: 10.1007/s12008-023-01288-7

[46] Straubinger A, Rothfeld R, Shamiyeh M, Büchter K-D, Kaiser J, Plötner KO. An overview of current research and developments in urban air mobility – Setting the scene for UAM introduction. Journal of Air Transport Management. 2020;**87**:101852. DOI: 10.1016/j.jairtraman.2020.101852

[47] Kuhn H, Falter C, Sizmann A. Renewable Energy Perspectives for Aviation. 2011

[48] Rezende R, Barros J, Perez V. General Aviation 2025 - A study for electric propulsion. 2018. DOI: 10.2514/6.2018-4900

[49] Gipson L. Advanced Air Mobility (AAM). NASA. 2019. Available from: http://www.nasa.gov/aam [accessed July 5, 2023]

[50] Bagassi S, De Crescenzio F, Lucchi F, Masotti N. Augmented and virtual reality in the airport control tower. 2016

[51] Gorbunov AL, Nechaev EE. Augmented reality Technologies in air Transport Control Systems. In: 2022 Systems of Signals Generating and Processing in the Field of on Board Communications. Moscow, Russia: IEEE; 2022. p. 1-5. DOI: 10.1109/IEEECONF53456.2022.9744399

[52] NASA Aeronautics Research Institute (NARI). NASA's Advanced Air Mobility. NARI. n.d. Available from: https://nari.arc.nasa.gov/aam-portal/ [accessed July 5, 2023]

[53] Hill BP, DeCarme D, Metcalfe M, Griffin C, Wiggins S, Metts C, et al. UAM Vision Concept of Operations (ConOps) UAM Maturity Level (UML) 4. Washington, DC, USA: NASA; 2020 Dec [cited 2023 Jul 5]. (NTRS - NASA Technical Reports Server). Report No.: 20205011091. Available from: https://ntrs.nasa.gov/citations/20205011091

[54] Bassey R. Vertiport Design. Washington, DC, USA: Federal Aviation Administration. Sep 2022 [cited 2023 Jul 5]. Report No: 105. Available from: https://www.faa.gov/sites/faa.gov/files/eb-105-vertiports.pdf

[55] Fontaine P. Concept of Operations Version 2.0. Washington, DC, USA: Federal Aviation Administration; Apr 2023. [cited 2023 Jul 5]. (Urban Air Mobility (UAM)). Available from: https://www.faa.gov/sites/faa.gov/files/Urban%20Air%20Mobility%20%28UAM%29%20Concept%20of%20Operations%202.0_0.pdf

[56] Gipson L. Advanced air mobility project. NASA 2020. http://www.nasa.gov/aeroresearch/programs/iasp/aam/description [accessed July 5, 2023]

[57] Rostami M, Kamoonpuri J, Pradhan P, Bardin J, Oyama Y, Choe A, et al. Enhancing Urban Air Mobility Development Using Extended Reality Technology. Conference presentation presented at: Canada-Korea Conference on Science and Technology; Niagara Falls, ON, Canada; Jul 2022

[58] Wang Z, Zhang R, Lei Z, Descorme C, Wong M. New opportunities and challenges in energy and environmental catalysis (EEST2018). Catalysis Today. 2019;**339**. DOI: 10.1016/j.cattod.2019.08.001

[59] Wang Z, Bai X, Zhang S, Billinghurst M, He W, Wang Y, et al. The role of user-centered AR instruction in improving novice spatial cognition in a high-precision procedural task. Advanced Engineering Informatics. 2021;**47**:101250. DOI: 10.1016/j.aei.2021.101250

[60] Cohen Y, Naseraldin H, Chaudhuri A, Pilati F. Assembly systems in industry 4.0 era: A road map to understand assembly 4.0. International Journal of Advanced Manufacturing Technology. 2019;**105**:4037-4054. DOI: 10.1007/s00170-019-04203-1

[61] King H. Paris Air Show 2023: Airbus unveils immersive collaboration concept for cabin definition. FINN - Aviat Ind Hub FINN. 2023. Available from: https://www.wearefinn.com/topics/posts/pas-2023-airbus-unveils-immersive-collaboration-concept-for-cabin-definition/ [accessed July 5, 2023]

[62] Yuan ML, Ong SK, Nee AYC. Assembly Guidance in Augmented Reality Environments Using a Virtual Interactive Tool. Queenstown, Singapore: Singapore-MIT Alliance (SMA); 2005 Jan [cited 2023 Jul 5]. (Innovation in Manufacturing Systems and Technology (IMST)). Available from: https://dspace.mit.edu/handle/1721.1/7442

[63] Liu C, Cao S, Tse W, Xu X. Augmented reality-assisted intelligent window for cyber-physical machine tools. Journal of Manufacturing Systems. 2017;**44**:280-286. DOI: 10.1016/j.jmsy.2017.04.008

[64] Zhou F, Duh HBL, Billinghurst M. Trends in augmented reality tracking, interaction and display: A review of ten years of ISMAR. In: 2008 7th IEEE/ACM International Symposium on Mixed and Augmented Reality. Cambridge: IEEE; 2008. pp. 193-202. DOI: 10.1109/ISMAR.2008.4637362

[65] Wang Z, Bai X, Zhang S, He W, Zhang X, Zhang L, et al. Information-level AR instruction: A novel assembly guidance information representation assisting user cognition. International Journal of Advanced Manufacturing Technology. 2020;**106**:603-626. DOI: 10.1007/s00170-019-04538-9

[66] Henderson SJ, Feiner SK. Augmented reality in the psychomotor phase of a procedural task. In: 2011 10th IEEE International Symposium on Mixed and Augmented Reality. Basel, Switzerland: IEEE; 2011. pp. 191-200. DOI: 10.1109/ISMAR.2011.6092386

[67] Syberfeldt A, Danielsson O, Holm M, Wang L. Visual assembling guidance using augmented reality. Procedia Manufacturing. 2015;**1**:98-109. DOI: 10.1016/j.promfg.2015.09.068

[68] Westerfield G, Mitrovic A, Billinghurst M. Intelligent augmented reality training for motherboard assembly. International Journal of Artificial Intelligence in Education. 2015;**25**:157-172. DOI: 10.1007/s40593-014-0032-x

[69] Fiorentino M, Monno G, Uva AE. Tangible digital master for product lifecycle management in augmented

reality. International Journal on Interactive Design and Manufacturing (IJIDeM). 2009;**3**:121-129. DOI: 10.1007/s12008-009-0062-z

[70] Yuan M, Ong SK, Nee A. Augmented reality for assembly guidance using a virtual interactive tool. International Journal of Production Research. 2008;**46**:1745-1767. DOI: 10.1080/00207540600972935

[71] Wang X, Ong SK, Nee AYC. Multi-modal augmented-reality assembly guidance based on bare-hand interface. Advanced Engineering Informatics. 2016;**30**:406-421. DOI: 10.1016/j.aei.2016.05.004

[72] Wang Z, Bai X, Zhang S, Wang Y, Han S, Zhang X, et al. User-oriented AR assembly guideline: A new classification method of assembly instruction for user cognition. International Journal of Advanced Manufacturing Technology. 2021;**112**:41-59. DOI: 10.1007/s00170-020-06291-w

[73] Büttner S, Mucha H, Funk M, Kosch T, Aehnelt M, Robert S, et al. The design space of augmented and virtual reality applications for assistive environments in manufacturing: A visual approach. In: Proc. 10th Int. Conf. PErvasive Technol. Relat. Assist. Environ. New York, NY, USA: Association for Computing Machinery; 2017. pp. 433-440. DOI: 10.1145/3056540.3076193

[74] Ong SK, Wang ZB. Augmented assembly technologies based on 3D bare-hand interaction. CIRP Annals. 2011;**60**:1-4. DOI: 10.1016/j.cirp.2011.03.001

[75] Dori D, Tombre K. From engineering drawings to 3D cad models: Are we ready now? Computer-Aided Design. 1995;**27**:243-254. DOI: 10.1016/0010-4485(95)91134-7

[76] Pilkaite T, Nenorta V. Digital product definition data practices. In: BALTGRAF Sel. Pap. Int. Conf. Eng. Graph. Vol. 12. Vilnius, Lithuania: Vilnius Gediminas Technical University, Department of Construction Economics & Property; 2013. pp. 171-176

[77] Alemanni M, Destefanis F, Vezzetti E. Model-based definition design in the product lifecycle management scenario. International Journal of Advanced Manufacturing Technology. 2011;**52**:1-14. DOI: 10.1007/s00170-010-2699-y

[78] Liu F, Qiao LH. Product information Modeling and organization with MBD. Applied Mechanics and Materials. 2012;**163**:221-225. DOI: 10.4028/www.scientific.net/AMM.163.221

[79] Barfield W, Caudell T, editors. Boeing's Wire Bundle Assembly Project. In: Fundamentals of Wearable Computers and Augmented Reality. 1st ed. Boca Raton: CRC Press; 2001. pp. 462-482 [cited 2023 Jul 5]. Available from: https://www.taylorfrancis.com/books/9780585383590/ DOI: 10.1201/9780585383590-21

[80] Sharma KJ, Bowonder B. The making of Boeing 777: A case study in concurrent engineering. International Journal of Manufacturing Technology and Management. 2004;**6**:254-264

[81] Shen Y, Ong SK, Nee AYC. Augmented reality for collaborative product design and development. Design Studies. 2010;**31**:118-145. DOI: 10.1016/j.destud.2009.11.001

[82] Nee AYC, Ong SK, Chryssolouris G, Mourtzis D. Augmented reality applications in design and manufacturing. CIRP Annals. 2012;**61**:657-679. DOI: 10.1016/j.cirp.2012.05.010

[83] Rekimoto J. Transvision: A hand-held augmented reality system for collaborative design. In: Proceedings of Virtual Systems and Multi-Media (VSMM '96). Washington, DC, USA: IEEE Computer Society; 1996. DOI: 10.5555/846220

[84] Huang Z, Li W, Hui P, Peylo C. CloudRidAR: A cloud-based architecture for mobile augmented reality. In: Proc. 2014 Workshop Mob. Augment. Real. Robot. Technol.-Based Syst. New York, NY, USA: Association for Computing Machinery; 2014. pp. 29-34. DOI: 10.1145/2609829.2609832

[85] Mourtzis D, Zogopoulos V, Vlachou E. Augmented reality application to support remote maintenance as a Service in the Robotics Industry. Procedia CIRP. 2017;**63**:46-51. DOI: 10.1016/j.procir.2017.03.154

[86] Insaurralde CC. Artificial intelligence engineering for aerospace applications. In: 2020 AIAA/IEEE 39th Digital Avionics Systems Conference (DASC). San Antonio, TX, USA: IEEE; 2020. p. 1-7. DOI: 10.1109/DASC50938.2020.9256770

[87] Brunton SL, Nathan Kutz J, Manohar K, Aravkin AY, Morgansen K, Klemisch J, et al. Data-driven aerospace engineering: Reframing the industry with machine learning. AIAA Journal. 2021;**59**:2820-2847. DOI: 10.2514/1.J060131

[88] Jordan MI, Mitchell TM. Machine learning: Trends, perspectives, and prospects. Science. 2015;**349**:255-260. DOI: 10.1126/science.aaa8415

[89] Wu X, Kumar V, Ross Quinlan J, Ghosh J, Yang Q, Motoda H, et al. Top 10 algorithms in data mining. Knowledge and Information Systems. 2008;**14**:1-37. DOI: 10.1007/s10115-007-0114-2

[90] Lynch C. How do your data grow? Nature. 2008;**455**:28-29. DOI: 10.1038/455028a

[91] Swart W. The Industrial Metaverse is "basically" a Digital Twin. Johannesburg, Gauteng, South Africa: 4 Sight; 2023 [cited 2023 Jul 5]. Available from: https://4sight.cloud/images/media/documents/01/785/metaverseblogpostwilhelmjanuary2022.pdf

[92] Brunton SL, Kutz JN. Data-Driven Science and Engineering: Machine Learning, Dynamical Systems, and Control. Cambridge: Cambridge University Press; 2019

[93] Batra R, Song L, Ramprasad R. Emerging materials intelligence ecosystems propelled by machine learning. Nature Reviews Materials. 2021;**6**:655-678. DOI: 10.1038/s41578-020-00255-y

[94] Lam TM, Boschloo HW, Mulder M, van Paassen MM. Artificial force field for haptic feedback in UAV teleoperation. IEEE Trans Syst Man Cybern - Part Syst Hum. 2009;**39**:1316-1330. DOI: 10.1109/TSMCA.2009.2028239

[95] Huang F, Gillespie RB, Kuo A. Haptic feedback and human performance in a dynamic task. In: Proceedings 10th Symposium on Haptic Interfaces for Virtual Environment and Teleoperator Systems HAPTICS 2002. Orlando, FL, USA: IEEE; 2002. p. 24-31. DOI: 10.1109/HAPTIC.2002.998937

[96] Srinivasan M. Haptic Interfaces. In: Durlach NI, Mavor AS, editors. Virtual Reality: Scientific and Technical Challenges. Washington, DC, USA: National Research Council, National Academy Press; 1995 [cited 2023 Jul 5]. p. 161-467. (Report of the Committee on Virtual Reality Research and Development). DOI: 10.17226/4761

[97] Arbeláez JC, Viganò R, Osorio-Gómez G. Haptic augmented reality (HapticAR) for assembly

guidance. International Journal on Interactive Design and Manufacturing (IJIDeM). 2019;**13**:673-687. DOI: 10.1007/s12008-019-00532-3

[98] Wang Z, Bai X, Zhang S, Billinghurst M, He W, Wang P, et al. A comprehensive review of augmented reality-based instruction in manual assembly, training and repair. Robotics and Computer-Integrated Manufacturing. 2022;**78**:102407. DOI: 10.1016/j.rcim.2022.102407

[99] Yasuda YDV, Cappabianco FAM, Martins LEG, Gripp JAB. Aircraft visual inspection: A systematic literature review. Computers in Industry. 2022;**141**:103695. DOI: 10.1016/j.compind.2022.103695

[100] Doğru A, Bouarfa S, Arizar R, Aydoğan R. Using convolutional neural networks to automate aircraft maintenance visual inspection. Aerospace. 2020;**7**:171. DOI: 10.3390/aerospace7120171

[101] Computer vision in aircraft maintenance and its transformative power — zeroG. 2021

[102] Lufthansa Technik Digital Summit on 21 April 2021 - Digital Customer Experience Center n.d.. Available from: https://www.lufthansa-technik-broadcast.com/lufthansa-technik-digital-summit-on-21-april-2021.html [accessed July 5, 2023]

[103] Hsu J. Boeing Q&A: Machine learning and AR-powered aircraft inspection. 2022. Available from: https://www.edge-ai-vision.com/2022/01/boeing-qa-machine-learning-and-ar-powered-aircraft-inspection/ [Accessed July 5, 2023]

[104] How computer vision based ATTOL system helps air crafts in landing and takeoff. Labellerr. 2022. Available from: https://www.labellerr.com/blog/how-computer-vision-based-attol-system-helps-air-crafts-in-landing-takeoff/ [accessed July 5, 2023]

[105] How Wayfinder is Using Neural Networks for Vision-Based Autonomous Landing... n.d. Available from: https://acubed.airbus.com/blog/wayfinder/how-wayfinder-is-using-neural-networks-for-vision-based-autonomous-landing/ [accessed July 5, 2023]

[106] Ham H, Wesley J, Hendra H. Computer vision based 3D reconstruction : A review. International Journal of Electrical and Computer Engineering. 2019;**9**:2394. DOI: 10.11591/ijece.v9i4.pp2394-2402

[107] Aldao E, González-Jorge H, Pérez JA. Metrological comparison of LiDAR and photogrammetric systems for deformation monitoring of aerospace parts. Measurement. 2021;**174**:109037. DOI: 10.1016/j.measurement.2021.109037

[108] Ardestani SM, Jin PJ, Volkmann O, Gong J, Zhou Z, Feeley C. 3D Accident Site Reconstruction Using Unmanned Aerial Vehicles (UAV). In: TRB 95th Annual Meeting Compendium of Papers. Washington, DC, USA: The National Academies of Sciences, Engineering, and Medicine; 2016 [cited 2023 Jul 5]. Available from: https://trid.trb.org/view/1394121

[109] What Is SLAM (Simultaneous Localization and Mapping) – MATLAB & Simulink. n.d. Available from: https://www.mathworks.com/discovery/slam.html [accessed July 5, 2023]

Section 2

Augmented Reality for the Construction Industry and Architecture

Chapter 5

Augmented Reality Application Areas for the Architecture, Engineering, and Construction Industry

Sara Rankohi, Mahsa Rezvani, Lloyd Waugh and Zhen Lei

Abstract

Augmented reality (AR) is among the technologies that have the potential to advance the Architecture, Engineering, and Construction (AEC) industry. Yet, studies show that there remain challenges in applying AR in AEC. According to the literature, the use of AR is focused on the construction phase to address performance, supervisory, and safety-related concerns. However, other phases of AEC projects could also benefit from this technology. Accordingly, this chapter provides an application-centric study to assess the state-of-the-art applications areas of AR in the AEC industry. Various applications have been identified as visualization and simulation; *in-situ* experience; real-time information retrieval; maintenance, inspection, and repair; project documentation; heavy equipment operation; educational training; health and safety; site navigation; and automated measurements. To further explore these application areas, a case study was conducted using the AR solution of Trimble XR10 with HoloLens 2 in a precast construction context. The results show that existing AR technologies and systems for simulation/visualization and construction quality control are still immature. The study highlighted the current use cases, the potential for technology improvements, and the obstacles that hinder the widespread AR implementation in the AEC industry. Considering these factors, further directions and future research paths for innovators are proposed.

Keywords: augmented reality, AEC, construction, state-of-the-art, construction management

1. Introduction

The concept of augmented reality (AR) can be traced back to the 1960s when researchers began exploring ways to create immersive experiences through computer-generated environments [1]. In 1968, Ivan Sutherland, the developer of the first head-mounted display (HMD), proposed the concept of "the Ultimate Display," which included the idea of overlaying computer-generated graphics onto the real world. This can be considered an early precursor to augmented reality. The term "augmented reality" itself was coined in the early 1990s by Tom Caudell, a researcher at Boeing,

 IntechOpen

to describe a digital display system he was working on to assist in aircraft assembly. While AR has its roots in the 1960s, it wasn't until the 2000s that AR started to gain more attention and practical application, with advances in computer vision, wearable technology, and mobile devices [2–4].

Augmented reality (AR) is an emerging and interactive technology that can add contextual content to real life, compare the digital world with the real world, and fill the gaps between them. AR systems are generally composed of three elements, which are computing, data, and presentation [5]. By using this technology, the user can see the real world with superimposed digital and virtual data through the application of software such as SICURA and SMART, and hardware such as Microsoft HoloLens [6]. Through integrating cameras, HMDs, GPS sensors, and Internet access of smartphones and tablets, in mobile AR applications real-world environments are overlaid with context-based, dynamic, and interactive digital content [7]. This emerging technology is currently used in applications such as education, gaming and entertainment, manufacturing, medical fields, cultural heritage, scientific visualization, and military applications [2, 3]. Due to high interest in AR, currently, researchers and developers from various domains are working collaboratively to expand knowledge and develop new augmented reality applications that can provide significant benefits of AR technologies to various fields of research and development [6].

In this Chapter, first, the AR research context (the industry, audience, and complementary technologies) is described in Section 2. Next, the prototype applications and application opportunities are identified in Section 3. In Section 4, the authors candidly identify the current challenges facing AEC industry applications. Then, the drivers and opportunities that will enable AR applications in the AEC industry are optimistically described in Section 5. Following that, a case study that demonstrates opportunities and challenges is provided in Section 6.

2. The AEC industry

The AEC industry is a significant sector that plays a vital role in economic development of countries [7]. This industry is a groundbreaking domain incorporating big data, artificial intelligence, and a wide range of knowledge fields and businesses, involved in the design, site preparation, construction of supply lines, construction of infrastructure, off-site construction, construction installation and on-site assemblies, maintenance, and repairs [1, 6]. The AEC industry is experiencing a radical change as project participants are compelling project visibility to increase construction efficiency and reduce project risks. This has enhanced the use of new technologies throughout projects' lifecycle [8]. As technology becomes more mature and applicable, AR in turn will become an inestimable tool that can change the future of the AEC industry [7–9]. This is in accordance with the scholars which shows AR is recently gaining a momentum in the AEC industry due to the need for innovation in conducting construction-related activities.

This Chapter about AR is significant to AEC professionals, scholars, researchers, practitioners, the government, and other public agencies to expand their knowledge of how augmented reality technologies can be applied and adopted in the AEC industry. First, an application-centric literature review on the state-of-the-art AR technologies, application areas, barriers, and drivers in the AEC industry is described. Next, the results of a very recent case study are provided, which assess the state-of-the-art regarding the application of AR in AEC projects.

2.1 Industry sector and target audience

Various project types can gain advantage from AR technologies including:

- Infrastructure/public/municipal, e.g., evaluating dynamic city models and developing an model for transportation emission [10],
- Residential, e.g., virtual and augmented reality for designing and customizing mass housing [5],
- Building/commercial, e.g., visualizing high-rise building construction strategies [11], maintenance of exterior closures and interior finishes of walls and in buildings construction [12],
- Heavy/highway, e.g., developing virtual reality systems for optimized simulation of road design data [5], segmentation and recognition of highway assets using image-based 3D point clouds and semantic Texton forests [13], and
- Industrial, e.g., application areas for augmented reality in industrial construction [14].

Various target audiences can benefit from AR technologies, due to the complexity of AEC projects and the collaborative nature of this industry:

- Building systems engineers, e.g., electrical, mechanical, and structural engineers [5],
- Workers, e.g., machine operators, site supervisors, and technicians [1],
- Project end users, e.g., building occupants, residents, and office employees [2],
- Design teams, e.g., architects, engineers, and interior and exterior designers [3],
- Inspectors, e.g., project safety officers [12],
- Schedule and budget professionals, referred to as project managers [11],
- Engineering students [5],
- Other stakeholders, e.g., clients and building owners [5].

2.2 Integration with other information and communication tools and technologies

Overlaying computer graphics demonstrating object-related information to the user's field of view requires data related to spatial coordinates and practical constraints in that application field [6]. This indicates that AR cannot work independently unless all information is provided [6]. Studies show that to adopt AR effectively in AEC projects, it should be assisted by technologies like Building Information Modeling (BIM), Computer-Aided Design (CAD), Geographic Information System (GIS), and Global Positioning Systems (GPS) [1].

BIM: Mutual application of AR and BIM can significantly impact and benefit building projects from preconstruction to post-construction stages using other technological advancements including big data and wireless sensor technologies, gathering and applying information throughout the project lifecycle [6]. Yet, the AEC industry needs adaptation to gain full advantages of combining AR and BIM tools and technologies [6, 11–14].

CAD: Using CAD models, AR provides 3D images that augment the real world. Displaying images are projected directly through display devices, which requires a database that involves previously developed plans to support the AR application [6, 15, 16].

GIS: This can be used as a supporting tool to collect real-time data from the surrounding environment to be applied along with AR tools. Field-based GIS gives geo-referenced, topographic, and cartographic information, which is vital in employing virtual images in the real-world environment. A geo-referenced database is critical as it keeps the available data for analysis related to every real object of the physical space [6–8].

GPS: This can be used to acquire a user's location whenever they change their positions in an outdoor environment. With technological advancements, smaller, light-weighted, and stronger devices equipped with GPS for self-localization, digital camera lenses, and measures to transmit information at ample bandwidths are available, which can display real-world environments augmented with GPS/GIS data and 3D CAD models [6]. These tools together with AR are essential for achieving practical functions in enhancing performance [17].

3. Application areas

The application of AR in the AEC industry can impact project performance in different fields such as providing collaborative opportunities; enhancing visualizations and simulations; monitoring progress and comparing as-built with as-planned status of projects; enhancing virtual construction site visits; pre-empting work schedule disputes; planning construction activities; and developing 3D building models and training site workers for similar projects [6]. Further explanations about these areas are provided below.

3.1 Visualization and simulation

AR enhances users' visualization ability via adopting context-related objects, such as gazing through wall panels to see columns, or virtually walking under the ground to inspect the installation of subsurface utilities [6, 18]. AR can be also adopted to visualize the design and assist the design team during the design process [19]. In terms of simulation, AR provides authentic virtual models that can be used to simulate real-world environments. For instance, extracting site measures from 3D augmented models for evaluation of site progress, plan discrepancies, misplacements, and earthquake-induced building damage [5–7].

3.2 *In-situ* experience

Augmented reality (AR) can provide users with *in-situ* experiences such as providing virtual access, verifying the models, and giving warnings. *In-situ* experience can be developed by walking around a particular landscape. In developing such a virtual environment there is the possibility of adding abstract information (i.e., environmental noise contours around the building) into the augmented view [20]. The user can virtually examine the constructed structures to see whether they suitably fit in with the environment. This application can enhance the use of collaborative project delivery methods, such as integrated project delivery (IPD), in which early involvement of project participants is required [17]. Stakeholders' early engagement in the construction process can reduce costs due to changes [6, 20]. For instance, *in-situ* inspections of the site to verify the as-planned status of the project is alignment with the as-is situation [6, 21], and *in-situ* real-time live warnings given to workers to avoid unseen dangers in the area [6, 22].

3.3 Real-time information retrieval

Augmented reality (AR) can provide project participants with real-time access to project information at various stages of the project. For instance, site inspectors can use AR technologies on-site to conduct inspection activities. Mobile AR technologies can provide professionals with adequate information on projects needed for task management activities including visualizing task locations in a virtual environment [23–25]. A mobile BIM AR system with cloud storage capabilities can improve task efficiencies by enhancing the information retrieval process [14–16]. In the transportation domain, AR users can access live information such as car maintenance data, traffic incidents, and route changes, as well as visual/auditory information about the adopted route [6, 20].

3.4 Maintenance, site inspection, and repairs

Augmented reality (AR) can help site maintenance workers to avoid obscured objects such as buried electrical wiring and structural elements that affect the outdoor environments [6]. This accelerates maintenance and renovation operations and minimizes the amount of accidental damage that mostly happens during maintenance [21, 22]. The mounting technical complexity and a high degree of diversity of components installed turn out to be great hindrances for service technicians maintaining industrial facilities [26]. Utility inspection and maintenance activities can be assisted if AR can be used to visualize underground utilities. AR enables users to gaze under the ground and inspect the subsurface utilities. In an AR-assisted inspection method, user's normal experience is augmented with context-related or geo-referenced virtual objects [27].

3.5 Project documentation

Augmented reality (AR) can be used significantly to document the project's progress. With recent developments in the field of AR, users can directly augment 2D project drawings with generated 3D models. AR combined with BIM and 3D modeling software, can provide elaborated, interactive models of buildings to be shared with project stakeholders at early stages of the project [6]. Clients are permitted to visualize the realistic outcomes of the project and necessary changes can be made before the construction starts. All these models and other project documents can be stored in augmented reality database, which can later be accessed by all project parties [28].

3.6 Heavy equipment operation

Augmented reality (AR) provides the possibility to give training to service technicians and heavy equipment operators in the construction process [29].

Using this training platform, the experienced workers can execute complex operations better and avoid project time delays [30–32]. Real-time tracking capability of AR tools can enhance workers' awareness in hazardous situations at the job sites [6, 17, 18, 33]. Applying augmented models of heavy equipment, such as cranes and boom lifts, allows workers to use their headsets to rehearse operating heavy machinery in a safe virtual environment, and enhance their learning experience and development [6, 20].

3.7 Educational training

Augmented reality (AR) provides new opportunities to effectively train and educate students or novices with a higher level of cognition and fewer hazards [13]. Safety training in the construction industry is challenging and safety-related programs cost a lot of money and time [25]. Training scenarios can be provided to students with the aid of an AR tools and technologies. The virtual intuitive site-safety learning enhances students' awareness of safety while lowering downtime and training costs [6]. Students can easily see AR images on their mobile devices due to the rapid development of AR applications [34].

3.8 Health and safety

The AEC industry is associated with highly hazardous situations that can cause danger to site workers. Thus, safety and health are one of the highest priorities for laborers. AR emerging technologies can enhance workers' health and safety when applied correctly [35]. Vehicles equipped with AR devices can give road guidance and real-time data about dangerous site conditions to operators. Workers can use AR devices and cameras connected to their safety helmets, to see and hear through fire, smoke, heavy rain, flood, bad weather, and other dangerous conditions on-site [30–32, 34]. ICT tools integrated with AR can provide site workers with well-interpreted information to monitor the difference between standard safety requirements and unsafe site conditions [36].

3.9 Site navigation

Augmented reality (AR) coupled with GIS data with augmented visual landmarks in the real-world environment can lead users to a particular location or direction [6]. In the realm of construction site navigation activities, the data integrated into AR applications serves three primary purposes: (1) revealing obscured elements (such as obstructed objects or buried components), (2) visualizing forthcoming constructions (anticipating the future), and (3) perceiving imperceptible aspects (including site boundaries, organizational details, alignment information, or infrequent environmental incidents like rare floods) [6]. AR can assist site workers in navigating throughout the construction site or within a particular facility [20–24].

3.10 Automated measurements

AR can be used to extract measurements of the physical objects (width, height, and depth) on-site. This data can be integrated into 3D models to generate more accurate structural visualization of the project based on the project's ultimate geometry [6]. AR can enable field workers to conduct automated measurements on-site during the design phase. AR allows users to capture building structures accurately, which can be adopted for real-time building simulations [3–7]. When an AR device is worn, workers can tap and automatically make measurements of the built elements and compare them to the planned measurements. This allows workers to find discrepancies between as-planned and as-built models and swiftly adjust them to avoid higher costs and delays in the process [37]. Another use of AR is the ability to walk through real civil infrastructure while simultaneously viewing a virtual 3D model of the same infrastructure as recorded several years prior. This is particularly valuable as a means of detecting trends in the deterioration of concrete infrastructure [38].

4. Current challenges

Augmented reality (AR) has great potential to revolutionize the AEC industry by enhancing visualization, improving communication, and streamlining project management. However, there are several challenges that need to be addressed for successful implementation of AR. Here are some of the key challenges:

4.1 High initial cost and low return on investment

Augmented reality (AR) implementation in construction requires substantial investment in hardware, software, training, and maintenance. Construction companies need to carefully evaluate the cost–benefit ratio and calculate the potential return on investment. Demonstrating the tangible benefits and long-term cost savings of AR technology can be challenging but is essential for wider adoption. Since AEC projects are typically large and complex, a prevailing notion indicates that applying AR tools and technologies would require a significant number of expensive devices, such as HMDs and other related equipment [8, 16–18, 33]. Subsequently, a major initial investment would be necessary to integrate AR into construction and infrastructure projects, which can expand the overall project cost [8]. Thus, it is essential to conduct research and development activities toward developing low-cost AR tools and technologies suitable for the AEC industries [8].

4.2 Nascent technologies

Augmented reality (AR) technologies are in nascency in the AEC industry. It requires powerful hardware and infrastructure to deliver real-time, high-quality AR experiences. Construction sites are often remote and lack stable internet connectivity, making it challenging to provide the necessary infrastructure for AR implementation. Additionally, AEC projects demand a higher degree of precision, consistency, and efficacy. However, current AR devices struggle to control the extremely sophisticated 3D information models regularly used in AEC projects [8]. Originally, AR devices were designed for the entertainment industry, thus, they might lack the on-site capabilities necessary for the AEC industry [8]. In addition, due to the complexity of this technology and a lack of technical skills and awareness, AR adoption in the AEC industry is slow [8]. Regarding the accuracy and precision, AEC projects require precise measurements and alignments. More development in AR systems is needed to accurately overlay virtual elements onto the physical environment. Achieving high accuracy and precision in AR tracking and alignment remains a challenge, especially in dynamic construction environments [20–25].

4.3 Insufficient adoption demand

Studies show that the demand for AR in the AEC industry is currently low. Nonetheless, with the ongoing evolution of technology and the formulation of customized answers to address the distinct requirements of the AEC sector, there's an anticipation of heightened acceptance [8]. To foster its adoption, initiatives aimed at raising awareness should be launched, highlighting the benefits of AR technology within AEC. These advantages encompass enhancing project comprehension, optimizing cost-effectiveness, and enabling efficient training [8]. Furthermore, educational establishments can contribute significantly by enticing fresh expertise in the field through the introduction of programs that offer specialized training in AR [33].

4.4 Lack of experts and insufficient training

Introducing AR technology requires training and upskilling the construction workforce. There may be resistance to adopting new technologies due to a lack of awareness or skepticism. Overcoming the learning curve and effectively training workers to use AR tools can be a challenge. A limited number of individuals are pursuing careers in AR due to its immature stage of development, complexities in adoption, and absence of standardized implementations. This lack of established norms complicates the evaluation of the knowledge and skills possessed by those working with AR technologies. Moreover, the scarcity of experts in this field poses difficulties in gauging their grasp of the technology and effectively involving them in financially significant AEC projects. To tackle this challenge, universities should introduce advanced education programs centered on AR, thereby bolstering research and development endeavors, and enhancing the technology itself. Concurrently, AR companies could support research and development initiatives within universities, thereby facilitating this progress [39].

4.5 Poor user experiences

Extended utilization of AR devices like HMDs can result in motion sickness, queasiness, perspiration, headaches, and even vomiting among users [8]. This poses a significant impediment to the widespread adoption of AR technology. Given the intricate and time-intensive nature of AEC projects, prolonged engagement with these devices can prove distressing for most users [8]. To address these concerns, endeavors should be directed toward enhancing the design and development of AR devices, aiming to minimize discomfort and alleviate the adverse effects experienced by users [8]. Additionally, incorporating frequent intervals during extended usage and restricting the duration spent in the virtual environment can aid in diminishing the occurrence of motion sickness and other related discomforts [8].

In addition, construction sites are complex and constantly changing environments. AR systems need to adapt to changing conditions and safety regulations. Ensuring that AR technology does not compromise safety and can handle various environmental factors like dust, vibrations, and lighting conditions is a significant challenge [2–4].

4.6 Integration, adaptability, and data security

Integrating AR seamlessly into existing construction workflows is crucial for adoption. AR systems need to interface with existing project management software, BIM data, and other AEC tools. Ensuring compatibility and smooth integration can be a technical and logistical challenge. The AEC industry involves multiple stakeholders, each using different software and systems. Achieving standardization and interoperability among various AR platforms, software, and hardware devices is essential to enable seamless collaboration and data exchange. AR systems in construction involve capturing, processing, and sharing sensitive project data. Ensuring data security, protecting intellectual property, and addressing privacy concerns are crucial. AEC companies must have robust data security measures in place to protect against unauthorized access or data breaches [22–25].

Addressing the challenges in this section requires a collaborative effort between AEC companies, technology providers, regulators, and industry associations. As AR technology continues to evolve, these challenges are likely to be mitigated over time, leading to more widespread adoption and integration of AR in the AEC industry.

5. Drivers and opportunities

While various barriers impede the adoption of AR technologies in the AEC industry, several drivers can impact users' motivations to apply these technologies in practice. These drivers are discussed below.

5.1 Enhancing comprehension of projects

The integration of AR technologies into the AEC sector is predominantly motivated by the advantages it brings in augmenting project comprehension [8, 10]. Through the application of AR, AEC endeavors can be virtually simulated, or tangible surroundings can be enriched with digital data, leading to a more profound grasp and visualization of the project [6–8]. This establishes a risk-free domain for the project, enabling the identification and resolution of any associated issues within a virtual realm [6]. Consequently, the utilization of AR enables the observation of diverse project stages, ultimately culminating in an enhanced holistic comprehension and reduction of project-related risks [8–12].

5.2 Reducing project overall costs

Augmented reality (AR) technologies have the potential to reproduce AEC projects' overall costs using the recently introduced digital methods and approaches, such as digital twin proposed by Oke et al. [6]. By identifying and rectifying problems in the virtual environment at early stages of the project, and optimizing project designs, it becomes possible to economize expenses and diminish the likelihood of human errors [8]. The amalgamation of the physical setting and the digital realm simplifies the supervision of project tasks, enabling virtual enhancements to be incorporated and real-time observations to take place. This, in turn, enhances the project's efficiency, resulting in superior outcomes achieved at a reduced expense [1, 6, 8].

5.3 Effective training scenarios

Leveraging AR technologies for employee training involves simulating real-life situations, affording employees the opportunity to hone their skills within a secure and regulated digital environment [8]. This methodology has the potential to considerably elevate training excellence through the provision of practical encounters within authentic contexts, ultimately fostering a more efficacious professional atmosphere [5–8]. Furthermore, this expedited acquisition of new proficiencies results in noteworthy reductions in training duration, consequently curtailing expenses while concurrently elevating the caliber of training [8, 9, 39].

5.4 Reducing damage and maintenance costs

Establishing a virtual environment through AR enables the efficient monitoring and juxtaposition of project advancement [7, 8]. This strategy mitigates the prospect of harm and subsequent repairs, given the prior digital simulation of the project. Additionally, the application of AR can unearth the most effective method for executing the project, culminating in diminished developmental expenditures [8, 15]. Consequently, the incorporation of AR within the AEC sector holds the potential to curtail project risks and trim development costs [40].

5.5 Improving user experiences

Studies show that AR technologies offer a unique and immersive experience to users [8–14]. The capabilities of these technologies offer a substantial prospect for reshaping interactions within our surroundings. AR bestows upon users a world of fresh opportunities, enabling them to engage with inanimate entities, foster deeper connections with individuals and surroundings, and envision their desires with precision. Within the AEC sector, these technologies hold remarkable potential [8–10]. For example, the immersive experience, offered by AR, allows workers to simulate thoughtless or mistaken actions alongside their ensuing outcomes, thus improving their training and safety protocols [8, 41].

6. Case studies

As discussed in previous sections, AR has various potential applications in AEC projects. However, its usage in AEC projects has not been widely implemented yet. Our investigation of AR applications identified major drivers of and barriers to the adoption of AR technologies in the AEC industry, which need further investigation. This section provides case study results on the implementation of AR technologies to conduct quality control (QC) inspection tasks in a precast manufacturing plant in Canada and identifies major challenges and recommendations to address those challenges [42]. Similar to other AEC industry environments, this case study environment is particularly challenging due to the component size, precision requirements, worksite congestion, and inherent safety hazards. For the detailed information regarding this case study, you can refer to the Master thesis by ref. [42].

6.1 The application of AR in the Strescon plant

The case study is conducted in a precast concrete manufacturer, Strescon Limited, located in Saint John, New Brunswick, Canada. The plant produces a variety of products including structural and architectural panels, concrete pipes, catch basins, and bridge girders. A group of 11 quality control inspectors at Strescon were selected as the focus group in the case study. Each participant was trained to work with the AR systems and the researcher observed their performance to ensure that they can work with the features without assistance.

The Augmented Reality solution of Trimble XR10 with HoloLens 2 [43] (in short XR10) was leveraged in this case study. The Trimble AR solution was first released in 2019 including the Microsoft HoloLens 2 head-mounted display and a hardhat approved by the Canadian Standard Association (CSA). Trimble Connect for HoloLens 2 (TCH) is the software component developed for XR10 to provide an AR experience, particularly for AEC practitioners. The AR solution of XR10 incorporates features such as TCH Measure, TCH Explore, TCH To-Dos, and TCH Navigate. TCH Measure allows users to measure any two points in a model or any two points with one in the model and one in the real world. **Figure 1(a)** shows an example where the user is measuring the overall dimensions of a precast concrete panel in the virtual model. TCH Explore consists of Explore Visibility and Explore Info. The Visibility feature allows a user to turn the visibility of model components on and off and view only the desired components. The Info feature provides a list of properties for each individual component in the cast unit model. These properties were already incorporated in the building information model of each cast unit. **Figure 1(b)** shows a user looking at the properties of a lifter hook in a precast concrete panel using TCH Explore Info. The To-Do feature in the Tools menu allows a user to view work orders or manage Requests for Information (RFIs). **Figure 1(c)** presents a snapshot of TCH To-Do where a user is creating a To-Do to send to office users. Users can navigate through the model, create section boxes, rotate, and manipulate the model using the TCH Navigate feature. **Figure 1(d)** shows a snapshot of TCH Navigate where a user is using the section-box tool and navigating the virtual model to gain a better look at the components of the cast unit model.

6.2 The implementation of XR10 at Strescon

The study involved volunteer QC inspectors and focused on the manufacturing process of a large architectural wall panel. The first step is to align the building

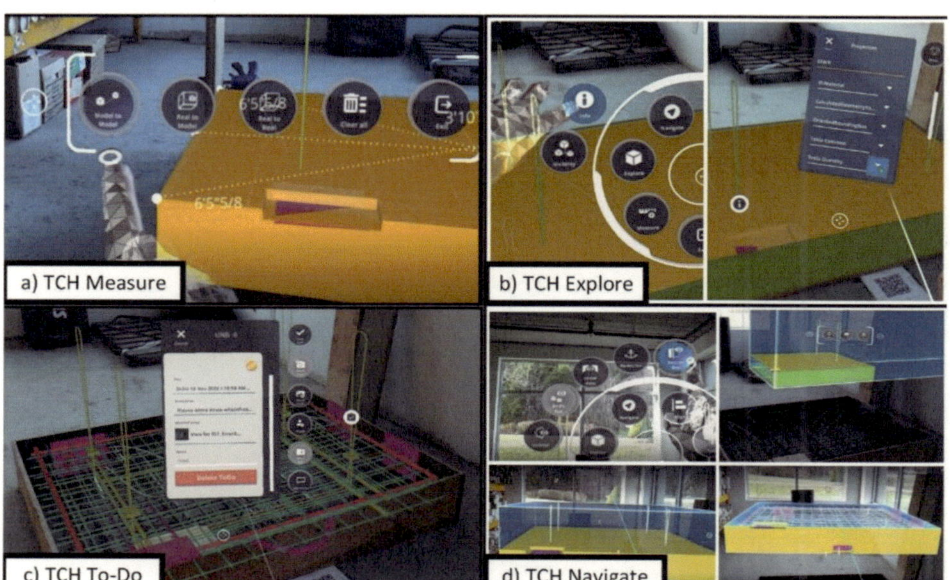

Figure 1.
Features of Trimble Connect for HoloLens 2 in an AR test on a precast concrete slab and its virtual model.

information model of the element with the cast unit. Trimble XR10 uses a marker-based spatial registration method to align models with their real-world counterparts. The proposed step-by-step instructions for model alignment is shown in **Figure 2**.

After aligning the virtual model with the real casting bed, the researcher observed three significant challenges in using XR10 for QC inspection and recommended solutions to facilitate the use of XR10 for QC inspection scenarios. These challenges and recommendations are summarized in sections below.

6.3 Challenges and recommendations

6.3.1 Transferring BIM models to Trimble cloud platform

Two distinct methods were identified for transferring the building information model from Tekla Structures to the Trimble Connect cloud platform: (1) one approach is to export the model of the building, in which the architectural wall panel (i.e., the test subject) is modeled, and (2) the second approach is to select the cast unit model of the architectural wall panel in the building model and export it individually and as a .ifc file.

The benefits and drawbacks that were observed in each approach are described below:

Approach 1: Exporting the entire building model as an .ifc file:

- Benefits: This approach eliminates the time to export each assembly unit of the building individually. Consequently, less workload is imposed on the BIM modeler when preparing .ifc files for use on the shop floor.

- Drawbacks: When a QC inspector attempts to inspect a cast unit assembly in the shop with XR10, they need to find the cast unit model that is to be inspected,

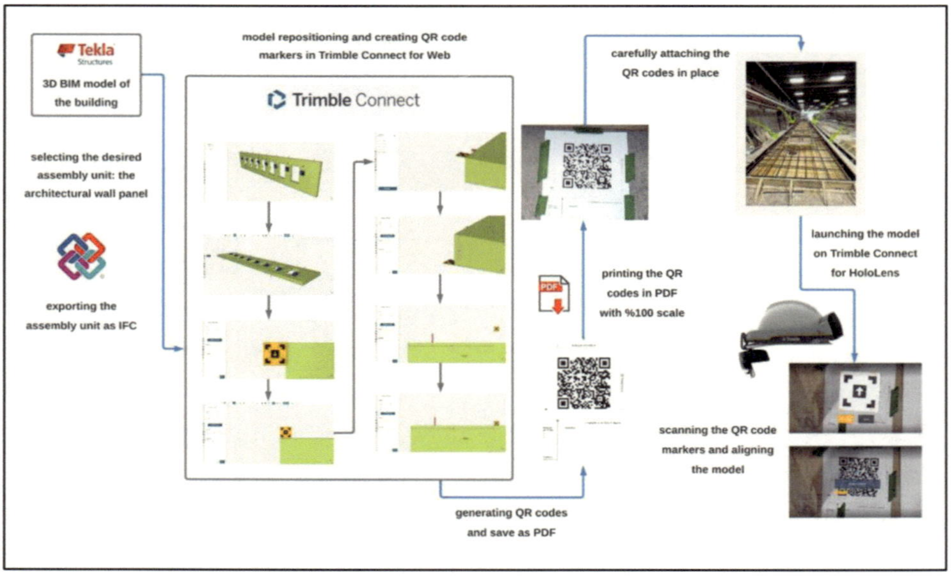

Figure 2.
The workflow of setting up the AR system in the Strescon shop floor.

from a long list of cast unit models that are modeled in a building. All other parts of the building model should be turned off for visibility, and the target cast unit model should remain on for visibility. This procedure imposes extra waiting time for a QC inspector, at the working table in the shop floor.

Approach 2: Exporting each cast unit model as a separate .ifc file:

- Benefits: The QC inspector is not required to find the desired cast unit from a list of all cast units in the model while wearing the XR10 in the dynamic environment of the shop floor. Therefore, the inspectors would start their inspection process by launching the .ifc file of each cast unit that is scheduled for inspection. Another benefit of this approach is that while some cast unit models such as wall panels are oriented vertically in the building model, the physical casting unit of that wall unit needs to be oriented horizontally on working tables in the shop floor. Therefore, the model needs to be repositioned in the Trimble Connect 3D viewer to match the orientation of the cast unit in the manufacturing line. However, if the model is not exported as a separate .ifc file and is exported along with other cast units in the building, repositioning the cast unit model imposes complication on the workflow of QC inspectors. When repositioning the model of a wall (from vertical view to horizontal view), a precast slab will also get reoriented to vertical view rather than horizontal view. In this case, a vertical orientation of a slab in the building model would not match the orientation of its counterpart cast unit in the manufacturing shop.

- Drawbacks: For large numbers of cast units that are scheduled for daily production, the BIM modeler needs to export each cast unit individually and send them out to the scheduling department for use by QC inspectors, which may impose extra time on the workflow of modeling and scheduling departments.

To avoid adding extra complication to the workflow of QC inspectors and minimize human errors when preparing cast unit model files for QC inspection, the second approach is recommended. **Figure 3** shows the repositioning of the architectural wall panel on the Trimble Connect Web platform.

6.3.2 Creating virtual QR code markers

The QR code markers need a clean surface area around the formwork for accurate scanning. Cover sheets are suggested to prevent marker dirtiness or damage to the QR code markers, see the location of QR code markers in **Figure 4**. Repositioning the QR code markers in the Trimble Connect platform was challenging, as it lacked options for snapping to points surrounding the model. As shown in **Figure 5**, a user may need to reposition the QR code markers several times. However, when the QR code marker is placed a distance from the cast unit model, there is no option for the user to snap to a point on the QR code marker and to verify the exact position of the marker relative to the cast unit. It is recommended that users manipulate coordinates of the QR code marker with great care to avoid errors. For BIM modelers it is also recommended to model bounding boxes around the cast unit models. The modeling of a bounding box in the 3D space around the cast unit model, allows a user to snap the QR code marker to any point within the bounding box. This method may help modelers verify the position of the QR code markers in a reliable manner.

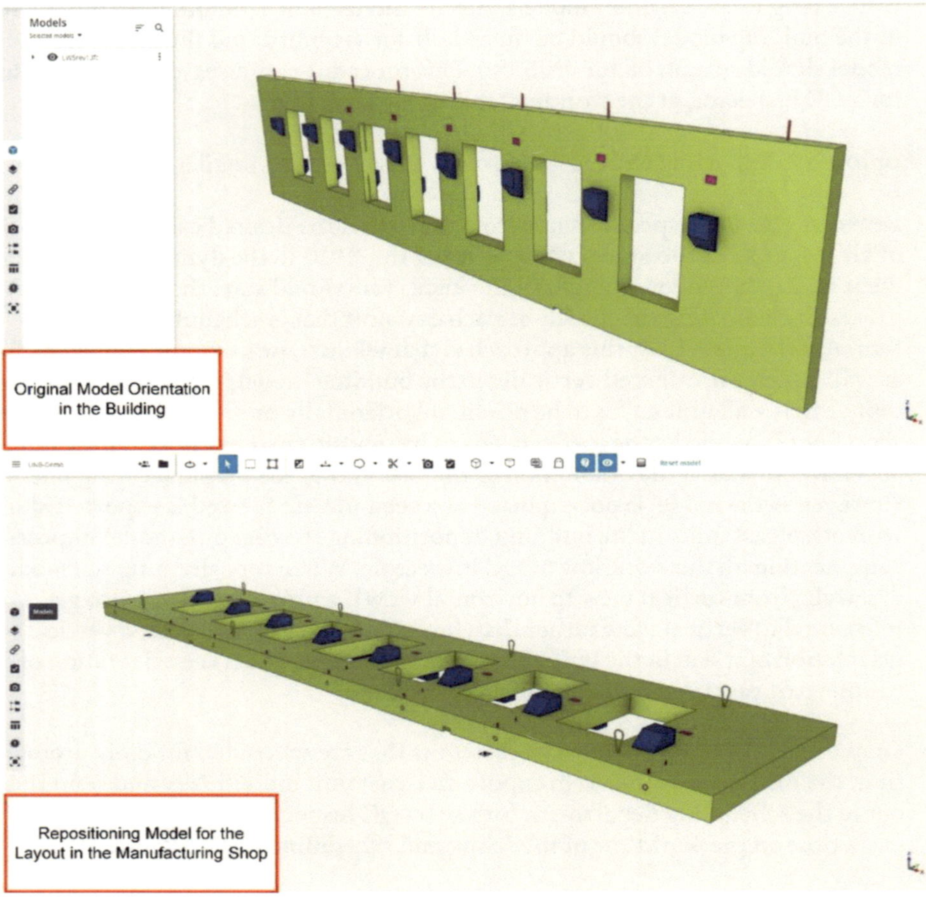

Figure 3.
Model repositioning on Trimble Connect for web.

Setting up the QR code markers on working tables posed another challenge. The use of formwork edges as reference points for marker placement could lead to misalignments between the 3D model and the as-built unit. To mitigate this issue, as shown in **Figure 6**, QC inspectors were advised to consider the discrepancies and consistently account for them during inspections.

6.3.3 Overall QC inspections

The use of the TCH Explore tool sometimes made it difficult to clearly see components in the real world because they were occluded by the superimposed virtual model. **Figure 7** shows the overall inspection of a cast unit where the user identified a misplaced component in the real precast unit based on the virtual model. Adjusting the opacity of the model is recommended to improve visibility during overall inspections. Snapping for measurements between the model and the formwork in the real cast unit was also challenging, with incorrect measurements resulting from

Figure 4.
Areas around the panel formwork for attaching QR code markers.

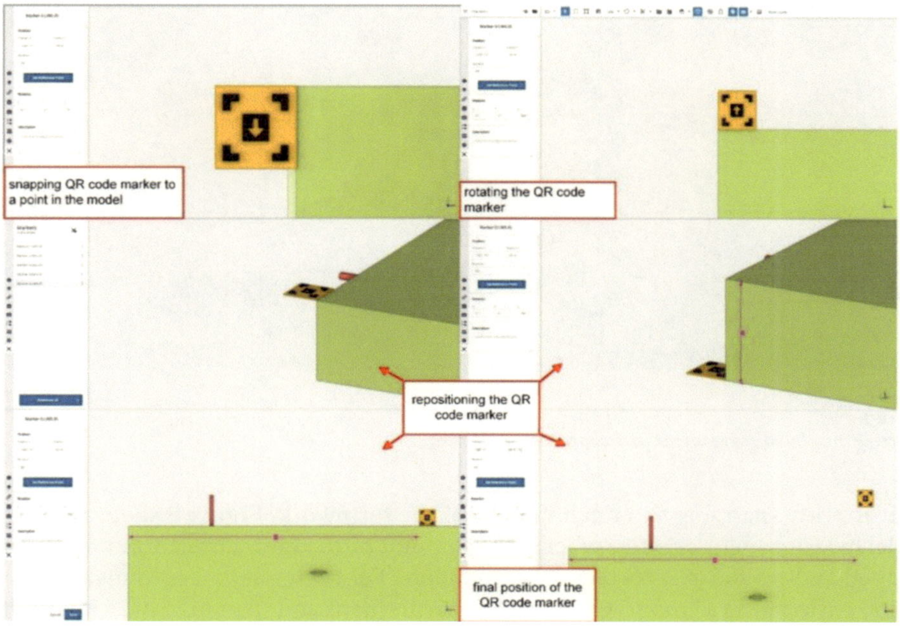

Figure 5.
Repositioning QR code markers on Trimble Connect for web.

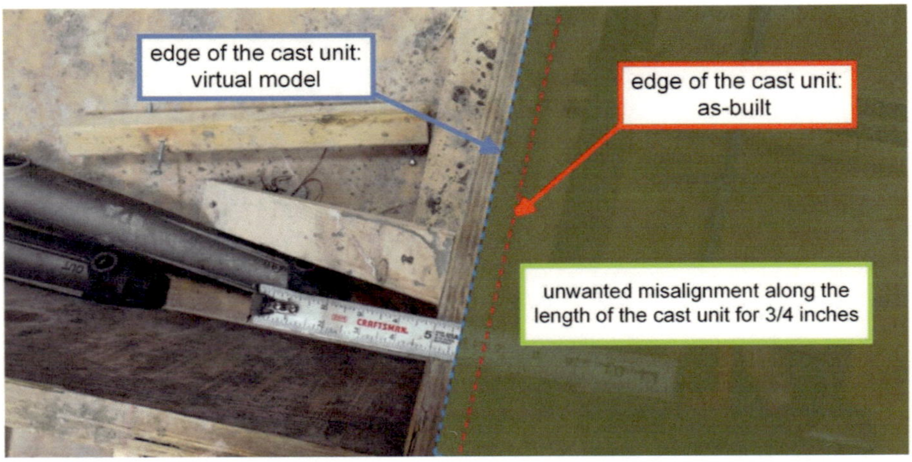

Figure 6.
Unwanted misalignment of the 3D model and the cast unit.

Figure 7.
Inspecting misplaced components in architectural wall panel.

inadvertently snapping to a rebar instead of the formwork. **Figure 8** shows a real-to-model measurement practice where the endpoint of the measurement is not correctly snapped and results in incorrect measurements. For tight tolerances of measurements, performing a "model-to-model" measurement and then matching it with its real measurement on the cast unit is advised to reduce errors.

Lastly, performing measurements in the depth of the formwork required the QC inspector to adopt uncomfortable positions. To overcome this, inspectors are recommended to relaunch the model from the TCH project browser, skip alignment with QR code markers, and view the model in 3D space for easier access. The TCH Navigate

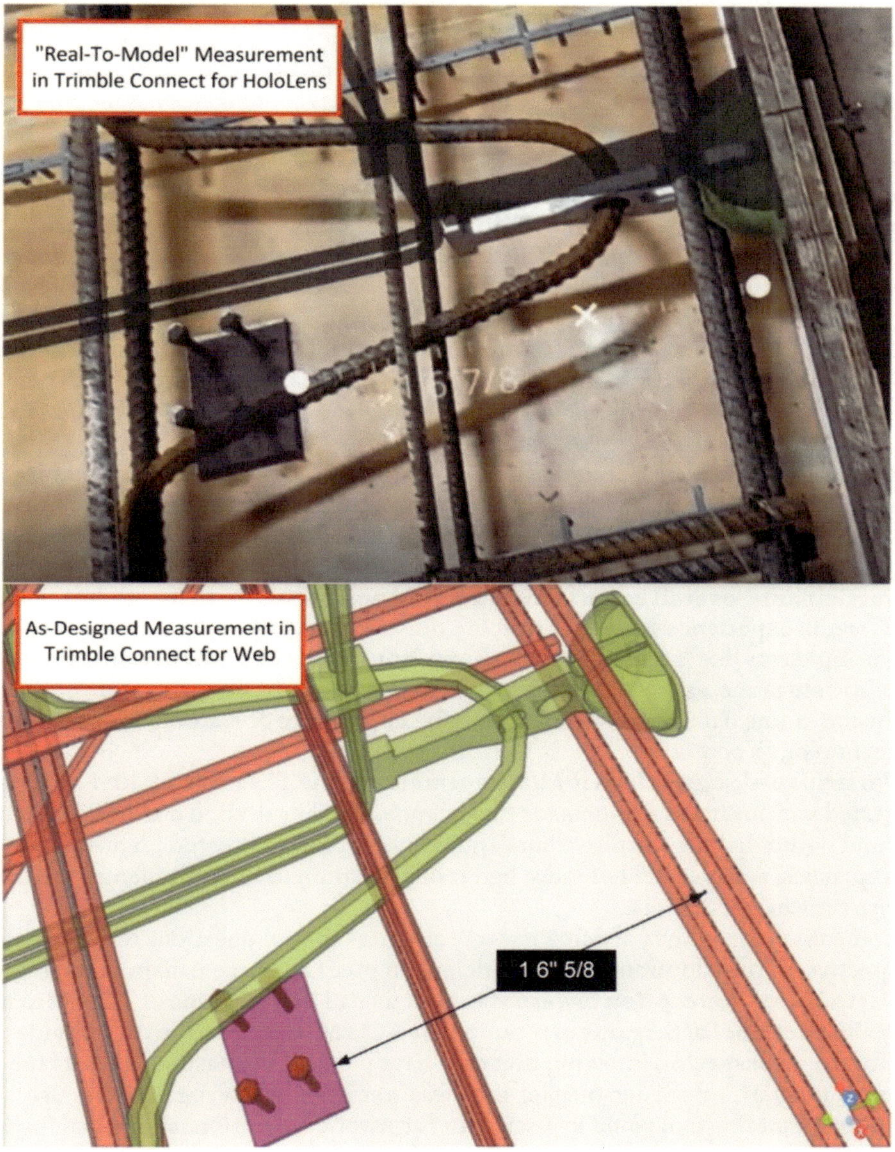

Figure 8.
Differences between the As-designed dimension (1′ 6″ 5/8) in the model and "real-to-model" measurement (1′ 5″ 7/8) in TCH.

tool can be used to enlarge the model and to perform "model-to-model" measurements conveniently.

6.4 Observations and findings from the implementation of XR10 at Strescon

In this section, researchers' observations on user experience and attitudes toward the TCH applications are provided. After implementing several applications of AR at the plant, Strescon QC inspectors (subsequently referred to as Participants) who were involved with the applications were asked a set of open-ended questions. These questions were designed to understand the Participants' experience with the Trimble AR solution and their perspective toward implementation of AR in precast concrete manufacturing. Participants' responses were recorded and transcribed. A content analysis method was carried out on the responses to identify similar codes and expressions. The results of Participants' experience with the TCH application and XR10 are summarized below:

Participants' experience and attitudes toward TCH Measure: Participants found the TCH Measure tool relatively hard to use compared to other features. The main reasons mentioned were difficulties in measuring irregular shapes, challenges in selecting tiny virtual objects like mesh and rebar, and problems with the TCH hand-gesture control.

Participants' experience and attitudes toward TCH Explore: Participants found the TCH Explore (visibility) and TCH Explore (Info) features relatively easier to use. They highlighted the intuitive layout and easy navigation of the application, as well as the easy-to-learn nature of these features.

Participants' experience and attitudes toward TCH interface: Participants had mixed opinions about the layout of information and features on the Trimble Connect application. While some found it intuitive and user-friendly, others mentioned challenging navigation. Overall, participants provided positive feedback about the layout.

Participants' overall experience with TCH and XR10: Participants described their overall experience with the Trimble Connect application and XR10 as positive, using terms like "great," "amazing," and "fun." However, they also mentioned usability issues and expressed mixed feelings such as frustration, annoyance, and confusion. Some participants emphasized the importance of training to fully utilize the technology's potential.

Participants' suggestions for improvements in the TCH and XR10: Participants suggested various improvements for the technology. They desired a more comfortable and lighter helmet, intuitive hand-pointer controls, automated identification of discrepancies, a bigger field of view, better organization of measurements, and more precise measurement tools.

In the research study, participants were asked a series of questions regarding their perspective on the adoption of AR technology in precast concrete manufacturing.

Participants' perception toward the most useful applications of TCH Measure in QC inspection: Participants mentioned the TCH Measure feature can provide comfort to the inspector, knowing that they have physically measured with a tape measure and verified the accuracy of the measurements. A few participants also mentioned that this tool could be useful for formwork inspection, and initial inspection. Other participants also found that the AR technology with TCH and XR10 can be useful for checking the overall dimensions or the location of hardware components in a cast unit. Several participants also concluded that the TCH Measure tool would not be useful at all.

Participants' perception toward the most useful applications of TCH Explore in QC inspection: Some participants mentioned that verifying the type and size of the components (e.g., rebar, welded plates, inserts, mesh, etc.), as well as finding out the location of components in the cast unit, are the most useful cases of using the TCH Explore tool. They also mentioned the "initial inspections" and the "final inspections" could also be the most useful cases of using TCH Explore but did not elaborate more on those. Another participant also mentioned the TCH Explore tool could be the most useful as a complementary tool to the conventional method of QC inspection to double-check and verify the inspections.

Participants' perception toward the most valuable features in the TCH application for QC inspection: Regarding the valuable features of the Trimble Connect application, participants mentioned tools like TCH Explore (Visibility), TCH Explore (Info), and TCH Measure. The TCH Explore (Visibility) tool was particularly appreciated by participants as it allowed them to isolate specific model components, enabling a focused inspection. The TCH Explore (Info) tool was mentioned by four participants, who found it helpful in quickly accessing information about model properties, saving time compared to conventional methods. Some participants also highlighted the usefulness of the TCH Measure tool, which eliminated the need for manual measurements with a tape measure. Additionally, participants found the overall 3D view of the model valuable, providing an immediate assessment of component placement and identification of any discrepancies.

Participants' perception toward the hindrances in implementing the AR application in precast concrete: When discussing hindrances or difficulties in using the Trimble Connect application with HoloLens in the production line, participants identified several concerns. The most common hindrance mentioned was the potential distraction caused by the busy and noisy shop environment, which could affect the QC inspector's focus. Safety was another frequently mentioned concern, as wearing the XR10 device might lead to reduced awareness of surroundings, especially in areas with equipment and debris. Participants also expressed the need for continuous learning and familiarity with the device, considering the learning curve as a hindrance. Other hindrances included durability in a dynamic construction environment, battery life, limited mobility, and the need for frequent adjustments to formwork setups.

Participants' perception toward the future of AR technology in precast concrete industry: Participants shared their expectations for the future of AR technology in precast concrete production, highlighting their overall perspective, required improvements in AR technology and Precast Concrete Manufacturing (PCM) processes, and potential benefits of implementing AR technology. While participants expressed optimism about the future adoption of AR technology in PCM, they also anticipated further development and refinement before full implementation. Improvements in AR technology mentioned by participants included enhanced accuracy in placing models on casting beds without recalibration, refined displays, improved precision, and preset options for model alignment. In terms of PCM processes, participants suggested simplifying QC tasks, expanding the level of detail in BIM software, and establishing efficient model alignment requirements. Anticipated benefits of AR technology implementation included facilitating preproduction meetings, improving efficiency and simplification of the QC process, and reducing errors.

Participants' perception toward the current use cases of AR technology in precast concrete industry: Participants identified specific areas within the QC inspection process where they believed AR technology could be effectively utilized. These areas included formwork inspections, initial and final inspections, supplementary inspection tools, and preproduction meetings. Participants were generally optimistic about using AR technology in initial and final inspections, considering it a supplementary method alongside 2D drawings. Additionally, they saw potential benefits in using AR technology during preproduction meetings to visualize products and identify issues.

Participants' suggestions for potential improvements in the AR technology for future of precast concrete: Regarding necessary improvements for using Trimble XR10 with HoloLens 2 in the QC inspection process, participants mentioned several areas that could be enhanced. These included user adoption, ambient light conditions in the shop and yard, maneuverability in the manufacturing line, organization of working tables, active notification systems for drawing revisions, and better time management. Concerns were raised about the impact of bright environments on model visualization, the need for sufficient space to move around while wearing XR10, and the importance of maintaining a clean and organized environment for effective use of the technology. Participants also emphasized the need for active notifications to inform individuals involved in QC inspection of drawing revisions. Moreover, participants mentioned the importance of allocating dedicated time for QC inspection with AR technology due to the presence of various tasks.

The results of the research indicated that AR users in the construction industry have a good understanding of various features in the existing technologies and their applications. In the case study, various applications of the TCH and XR10 have been explored, and users' experiences, attitudes, and perceptions toward the AR technology were discussed. Users found that existing AR technologies and systems in construction domain are still immature and provided suggestions for improving the current state of technologies in the industry. In summary, the study highlighted the current use cases, the potential for improvements of the AR technologies in construction, and the obstacles that hinder the widespread implementation of these technologies in the AEC industry.

7. Future outlook

In previous sections, numerous factors were identified that impede the wide applications of AR systems in the AEC projects. One aspect to consider involves the early stage of development for this technology and the gradual progress of augmented reality within the construction phase. Another factor to contend with is the limited effectiveness of human interactions with these emerging technologies, which poses an additional challenge when utilizing augmented reality. Considering these factors, it becomes important to focus future research and operational strategies related to the implementation of AR technologies in the construction phase are categorized into two categories, including technology-oriented and human-oriented topics. These two categories are further discussed in this section for future research directions.

7.1 Technology-oriented development

This group includes suggestions to improve the research conditions related to AR technologies in AEC projects and the development path of this technology, as follows:

- Improving the use of 4D/5D/6D AR tools to enhance project scheduling and costs, improving monitoring of project progress, and reinforcing decision-making activities [44].

- Developing advanced AR tools to display the model with a higher level of development (LOD) leads to the representation of model elements more accurately [45].

- Focusing on the use of I4.0 technologies and cloud-based systems to achieve a real-time ability in transferring, processing, and applying changes between the BIM and AR 3D models [46].

- Testing and developing AR-based digitally-twined systems with a multifunctional capability to evaluate the effects of this technology in the construction phase of projects [47].

- Enhancing AR-based systems to adapt this technology to infrastructure projects such as bridges and tunnels, which have different construction and monitoring conditions.

7.2 User-oriented development

At the current stage of applying AR technologies in the AEC industry, several human-oriented concerns could be confronting to investigate and might require a more accurate evaluation. Accordingly, certain suggestions are put forward as follows:

- Providing sufficient training to AR users in the project, before applying the technology. This training is provided automatically by AR itself to minimize the needs, which requires further studies [46].

- Increasing the use of AR to educate students and professionals to increase interactions and familiarity with this technology between the academic environment with the AEC industry [26].

- Involving and informing the industry professionals about the features and capabilities of AR to increase its acceptance rate. Improving public acceptance can lead to an increase in the use of AR in projects [43, 47].

- Developing project delivery methods (i.e., IPD) and business models (i.e., vertical integration) that facilitate the implementation of AR technologies and other emerging digital tools to improve construction project management practices [5, 42, 48].

In summary, the authors are optimistic about the adoption of AR in the AEC industry and revolutionization of the industry by these technologies in the future. Although current AR applications tend to be under the control of a single project participant and in tightly controlled environments, it is expected that the huge potential of AR will become more evident as multiple participants adopt integrated AR technologies across multiple project phases.

8. Conclusion

This chapter provides a review of the state-of-the-art AR technologies and their applications in AEC projects. In the presented work, the authors have attempted to provide a deeper understanding of the applications of AR for the AEC industry. Ten application areas for AR technologies in the AEC industry have been identified as visualization and simulation; *in-situ* experience; real-time information retrieval; maintenance, inspection, and repair; project documentation; heavy equipment operation; educational training; health and safety; site navigation; and automated measurements. Following that, various challenges and drivers for application of AR in AEC projects were extracted and a case study about AR application in a manufacturing plant in Canada was provided. Eventually, according to the literature review and case study results, implementing AR in AEC projects faces problems, which shows that this technology is still immature and needs further investigation and efforts. Accordingly, the authors of this chapter have suggested a series of technology-oriented and user-oriented research and executive directions in the "future outlook" section.

Acknowledgements

The authors express their sincere appreciation for the financial support received from the New Brunswick Innovation Foundation (NBIF) Emerging Projects under the grant titled "Towards a Digital-twin-based Prefabricated Construction Environment Using Data-driven Analytics and Immersive Technologies," with reference number EP_2022_009 and application number EP-0000000112. The contribution of in-kind support from StresCon Limited in Saint John, New Brunswick, Canada is also gratefully acknowledged for providing the testing environment essential for conducting the research case study.

Conflict of interest

The authors declare no conflict of interest.

Data availability statement

Some or all data that support the findings of this study are available from the corresponding author upon reasonable request.

Author details

Sara Rankohi[1]*, Mahsa Rezvani[2], Lloyd Waugh[3] and Zhen Lei[2]

1 School of Business and Management, Université de Québec à Montréal, Québec, Canada

2 Department of Civil Engineering, Off-site Construction Research Centre (OCRC), University of New Brunswick, Fredericton, New Brunswick, Canada

3 Department of Civil Engineering, University of New Brunswick, Fredericton, New Brunswick, Canada

*Address all correspondence to: rankohi.sara@uqam.ca

References

[1] Maqsoom A, Zulqarnain M, Irfan M, Ullah F, Alqahtani FK, Khan KIA. Drivers of, and barriers to, the adoption of mixed reality in the construction industry of developing countries. Buildings. 2023;**13**:872. DOI: 10.3390/buildings13040872

[2] Rezvani M, Zhen L, Rankin J, Waugh L. Current and future trends of augmented and mixed reality technologies in construction. In: Proceedings of the CSCE 2022 Conference; 25-28 May 2022; Whistler, British Columbia, Canada. 2022. pp. 866-870

[3] Shi Y, Du J, Ahn CR, Ragan E. Impact assessment of reinforced learning methods on construction workers' fall risk behavior using virtual reality. Automation in Construction. 2019;**104**:197-214. DOI: 10.1016/j.autcon.2019.04.015

[4] Yin Y, Zheng P, Li C, Wang L. A state-of-the-art survey on augmented reality-assisted digital twin for futuristic human-centric industry transformation. Robotics and Computer-Integrated Manufacturing. 2023;**81**:102515

[5] Rankohi S, Waugh L. Review and analysis of augmented reality literature for construction industry. Visualization in Engineering. 2013;**1**(1):1-18. DOI: 10.1186/2213-7459-1-9

[6] Oke AE, Arowoiya VA. An analysis of the application areas of augmented reality technology in the construction industry. Smart and Sustainable Built Environment. 2022;**11**(4):1081-1098

[7] Sepasgozar SM. Digital twin and web-based virtual gaming technologies for online education: A case of construction management and engineering. Applied Sciences. 2020;**10**(13):4678

[8] Kolaei AZ, Hedayati E, Khanzadi M, Amiri GG. Challenges and opportunities of augmented reality during the construction phase. Automation in Construction. 2022;**143**:104586

[9] Sommerauer P, Muller O. Augmented reality for teaching and learning-a literature review on heretical and empirical foundation. In: ECLS. 2018. p. 31. Available from: https://aisel.aisnet.org/ecis2018_rp/31

[10] Suk SJ, Ford G, Kang Y, Ahn YH. A study on the effect of the use of augmented reality on students' quantity take-off performance. In: ISARC Proceedings of the International Symposium on Automation, Robotics, and Construction; Vilnius. Proceedings of the 34rd ISARC; Taipei, Taiwan. Vol. 34. 2017

[11] Sutherland IE. A head-mounted three-dimensional display. In: Proceedings of the December 9-11, 1968, Fall Joint Computer Conference, Part I. 1968. pp. 757-764. DOI: 10.1145/1476589.1476686

[12] Teddlie C, Tashakkori A. Foundations of Mixed Methods Research: Integrating Quantitative and Qualitative Approached in the Social and Behavioral Sciences. London: SAGE; 2009

[13] Van Berlo L. C2B: Augmented reality on the construction site. In: 9th International Conference on Construction Applications of Virtual Reality. 2009. Available from: http://resolver.tudelft.nl/uuid:e7c27bbe-e1a3-4f46-b01b-df5d1437f273

[14] Kothari CR, Garg G. Research Methodology: Methods and Techniques. New Delhi: New Age International; 2014

[15] Li XX, Yi W, Chi HL, Wang X, Chan AP. A critical review of virtual and augmented reality (VR/AR) applications in construction safely. Automation in Construction. 2018;**86**:150-162. DOI: 10.1016/j.autcon.2017.11.003

[16] Martínez H, Skournetou D, Hyppölä J, Laukkanen S, Heikkilä A. Drivers and bottlenecks in the adoption of augmented reality applications. Journal of Multimedia Theory and Applications. 2014;**2**(1)

[17] Meza S, Turk Z, Dolenc M. Component-based engineering of a mobile BIM-based augmented reality system. Automation in Construction. 2014;**42**:1-12. DOI: 10.1016/j.autcon.2014.02.011

[18] Oke AE, Arowoiya VA. Evaluation of internet of things (IoT) application areas for sustainable construction. Smart and Sustainable Built Environment. 2021;**10**(3):387-402

[19] Alkhamisi AAS, Monowar MM. Rise of augmented reality. Current and future application areas. International Journal of Interment and Distribution Systems. 2013;**1**(4):25-34. DOI: 10.4236/ijids.2013.14005

[20] Arowoiya VA, Oke AE, Aigbavboa CO, Aliu J. An appraisal of the adoption of the internet of things (IoT) elements for sustainable construction. Journal of Engineering, Design and Technology. 2020;**18**(5):1193-1208. DOI: 10.1108/JEDT-10-2019-0270

[21] Bashabsheh AK, Alzoubi HH, Ali MZ. The application of virtual reality technology in architectural pedagogy for building constructions. Alexandria Engineering Journal. 2019;**58**(2):713-723. DOI: 10.1016/j.aej.2019.06.002

[22] Behzadan AH, Dong S, Kamat VR. Augmented reality visualization: A review of civil infrastructure system applications. Advanced Engineering Informatics. 2015;**29**(2):252-267. DOI: 10.1016/j.aei.2015.03.005

[23] Behzadi A. Using augmented and virtual reality technology in the construction industry. American Journal of Engineering Research. 2016;**5**(12):350-353

[24] Bergquist R, Stenbeck N. Using augmented reality to measure vertical surfaces [dissertation]. 2018. Available from: http://urn.kb.se/resolve?urn=urn:nbn:se:liu:diva-148868

[25] Chi HL, Kang SC, Wang X. Research trends and opportunities for augmented reality applications in architecture, engineering, and construction. Automation in Construction. 2013;**33**:116-122. DOI: 10.1016/j.autcon.2012.12.017

[26] Palmarini R, Erkoyuncu JA, Roy R, Torabmostaedi H. A systematic review of augmented reality applications in maintenance. Robotics and Computer-Integrated Manufacturing. 2018;**49**:215-228. DOI: 10.1016/j.rcim.2017.06.002

[27] Park CS, Kim HJ. A framework for construction safety management and visualization system. Automation in Construction. 2013;**33**:95-103. DOI: 10.1016/j.autcon.2012.09.012

[28] Wang M, Altaf MS, Al-Hussein M, Ma Y. Framework for an IoT-based shop floor material management system for panelized homebuilding. International Journal of Construction Management. 2018;**20**(2):130-145

[29] Wu W, Hartless J, Tesei A, Gunji V, Ayer S, London J. Design assessment in virtual and mixed reality environments: Comparison of novices and experts. Journal of Construction Engineering and Management. 2019;**145**(9):456-478. 04019049

[30] Zitzman L. Augmented Reality in Construction: 6 Applications in 2019. 2019. Available from: https://www.bigrentz.com [Accessed: August, 2019]

[31] Zollmann S, Hoppe C, Kluckner S, Poglitsch C, Bischof H, Reitmayr G. Augmented reality for construction site mentoring and documentation. Proceedings of the IEEE. 2013;**102**(2):137-154. DOI: 10.1109/JPROC.2013.2294314

[32] Chu M, Matthews J, Love PE. Integrating mobile building information modeling and augmented reality systems: An experimental study. Automation in Construction. 2018;**85**:305-316. DOI: 10.1016/j.autcon.2017.10.032

[33] Ogwueleka AC, Ikediashi DI. The future of BIM technologies in Africa: Prospects and challenges. In: Wu P, Li H, Wang X, editors. Integrated Building Information Modelling. Bentham Science Publishers; 2017. pp. 307-314

[34] Dakhil A. The contribution of the construction industry to economic development in Libya [C doctoral dissertation]. Liverpool, UK: Liverpool John Morres University; 2013. pp. 1-194. DOI: 10.24377/LJMU.t.00004454

[35] Danker F, Jones O. Combining augmented reality and building information modelling-an industry perspective on applications and future directions. In: Fusion-Proceedings of the 32nd eCAADe Conference; 10-12 September 2014; Newcastle upon Tyne, England. Vol. 2. 2014. pp. 525-536

[36] Dini G, Mura MD. Application of augmented reality techniques in through-life engineering services. Procedia CIRP. 2015;**38**:14-23. DOI: 10.1016/j.procir.2015.07.044

[37] Dunston PS, Shin DH. Key areas and issues for augmented reality applications on construction sites. In: Wang X, Schnabel MA, editors. Mixed Reality in Architecture, Design, and Construction. Dordrecht: Springer; 2009. DOI: 10.1007/978-1-4020-9088-2_10

[38] Cheng Z, Tang S, Liu H, Lei Z. Digital technologies in offsite and prefabricated construction: Theories and applications. Buildings. 2023;**13**(1):163

[39] Elghaish F, Matarneh S, Talebi S, Kagioglou M, Hosseini MR, Abrishami S. Toward digitalization in the construction industry with immersive and drones technologies: A critical literature review. Smart and Sustainable Built Environment. 2021;**10**(3):345-363. DOI: 10.1108/SASBE-06-2020-0077

[40] Gimeno J, Morillo P, Casas S, Fernandez M. An augmented reality (AR) CAD system at construction sites. In: Augmented Reality-Some Emerging Application Area. London, UK: Intech Open; 2011. pp. 15-32. DOI: 10.5772/26801

[41] Grubert J, Langlotz T, Zollmann S, Regenbrecht H. Towards pervasive augmented reality: Context-awareness in augmented reality. IEEE Transactions on Visualization and Computer Graphics. 2016;**23**(99):1. DOI: 10.1109/TVCG.2016.2543720

[42] Trimble XR10 with HoloLens 2 [Internet]. 2023. Available from: https://fieldtech.trimble.com/resources/mixed-reality/trimble-xr10-with-hololens-2inc.com/pricing/frames/content/solar_power.pdf [Accessed: July 25, 2023]

[43] Bradley C, Waugh L, Hanscom G. Advanced dam inspections with photogrammetric modelling. In: Proceedings of the 2019 Canadian Dam Association Conference; 6-10 October 2019; Calgary, Alberta. 2019

[44] Hammed A, Gantt JH Jr, Karim HA. Potential of mobile augmented reality for infrastructure field tasks. In: Proceedings Applications of Advanced Technology in Transportation Conf. Cambridge, MASS: AATT, ASCE; 2002. DOI: 10.1061/40632(245)54

[45] Issa RA, Shanbari H, Robey M, Blinn N. Using augmented reality to enhance construction management educational experiences. In: Proceedings of the 32nd CIB W78 Conference 2015; 27-29 October 2015; Eindhoven, The Netherlands. 2015. pp. 69-78. ISSN: 2706-6568. Available from: http://itc.scix.net/paper/w78-2015-paper-007

[46] Kamat VR, Golparvar-Fard M, Martinez JC, Pena-Mora F, Fischer M, Savarese S. CEC: Research in visualization techniques for field construction. In: Proceedings of the 2010 Construction; Sep 15-18; Michigan, US. 2010

[47] Matsika C, Zhou M. Factors affecting the adoption and use of AVR technology in higher and tertiary education. Technology in Society. 2021;**67**:101694

[48] Rezvani M. Mixed reality in precast concrete construction: A case study of quality control inspection tasks [Thesis]. Fredericton, New Brunswick, Canada: University of New Brunswick; 2023

Chapter 6

Augmented Reality in AEC Industry

Matevž Dolenc

Abstract

One of a kind products, processes, and partner groups typically characterise the architecture, engineering, and construction (AEC) industry. This has led to a challenging industry environment with islands of automation, difficult communication between stakeholders, and increasing specialisation. Building information modelling (BIM) is a modern approach to building design, documentation, delivery, and lifecycle management that addresses many AEC challenges. The outcome of this process is a comprehensive building information model that encompasses all relevant data and information related to the construction process as well as the built asset. This model is a primary resource for diverse end-user applications, including augmented reality, that could be essential in bridging the gap between the virtual and physical worlds and has become a component of modern digital workflows within the AEC industry. This chapter provides an overview of AEC specifics and how those reflect in implementing augmented reality in the context of BIM processes. It also delves into the challenges that must be addressed to facilitate the widespread adoption of augmented reality technology in the AEC sector.

Keywords: augmented reality, architecture, construction, engineering, AEC, computer-integrated engineering, building information modelling, BIM

1. Introduction

The architecture, engineering, and construction (AEC) industry is known for its complexity and multi-faceted nature, involving designing, constructing, and managing unique and intricate structures. However, this complexity also brings numerous challenges, including fragmented automation, communication barriers between stakeholders, increased specialisation, and a lack of integration between the various phases of the construction process. To address these challenges and increase industry efficiency and productivity, building data modelling (BIM) [1] has become a transformative approach.

BIM involves creating and managing a digital representation of a building, called the building data model, which contains comprehensive data and information about the construction process (**Figure 1**) [2]. This digital model is a central repository (Common Data Environment – CDE [3]) of information that all stakeholders can access and use, enabling better collaboration, improved decision-making and streamlined project delivery. However, traditional methods of interacting with the building information model, such as 2D drawings or computer screens, limit the ability of stakeholders to fully grasp and understand the complex spatial relationships and intricacies of the design [4, 5].

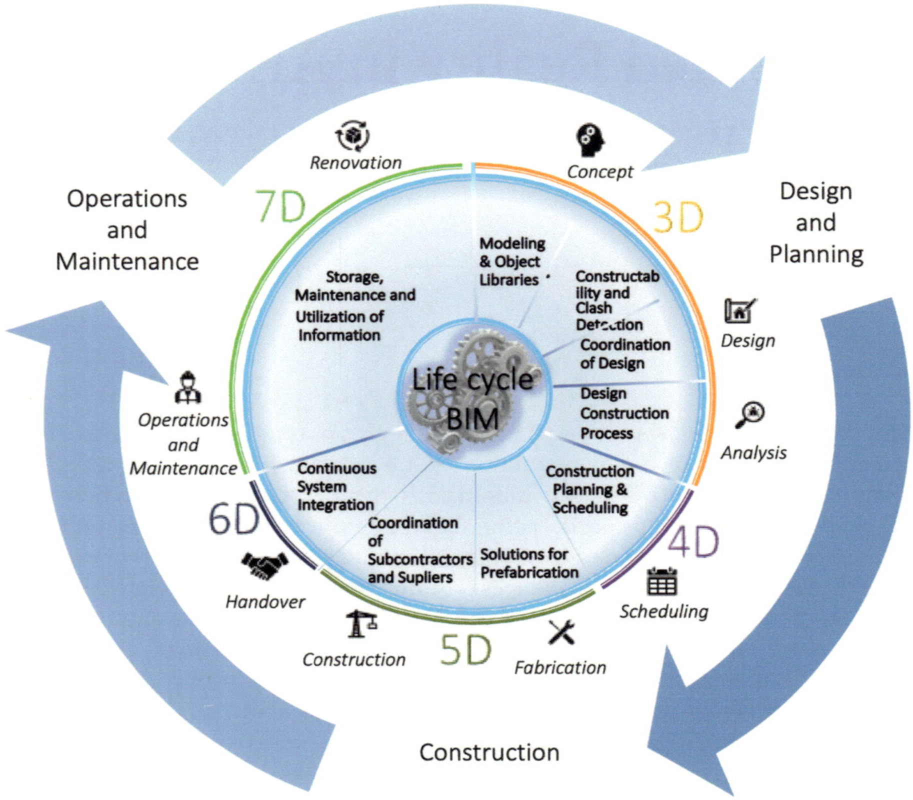

Figure 1.
BIM enables communication, collaboration, and visualisation throughout the building life cycle.

Augmented reality technology (AR) has garnered significant attention and demonstrated transformative potential across various industries, including the AEC sector [6, 7]. AR involves superimposing virtual information onto the natural environment, enriching the user's perception of the physical world by seamlessly integrating digital elements in real-time. In the AEC industry, AR empowers stakeholders to intuitively and immersively visualise and interact with building data models, effectively bridging the divide between the digital and real worlds (**Figure 2**).

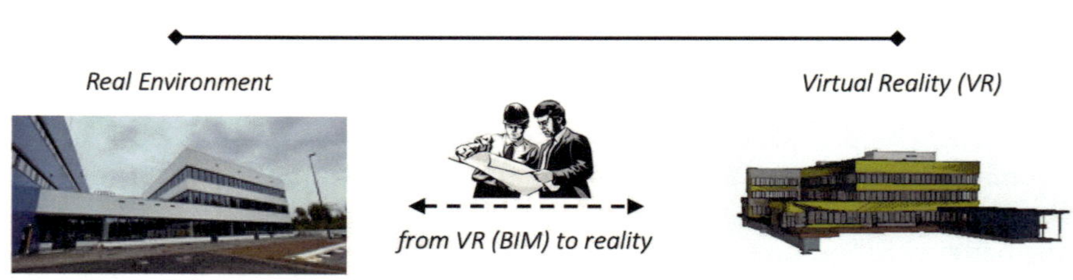

Figure 2.
Bridging the gap between virtual reality (BIM model) and the natural environment.

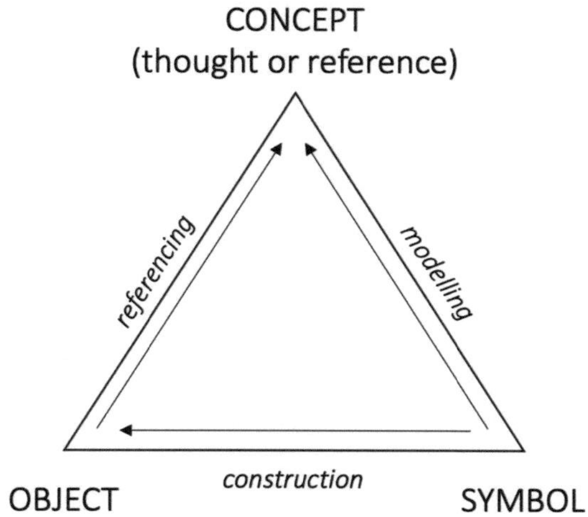

Figure 3.
The meaning triangle.

The integration of augmented reality into the AEC sector presents a multitude of advantages and opportunities [8]. The theoretical explanation of augmented reality's role can be illustrated through the concept of the meaning triangle (**Figure 3**) [9]. The concept refers to an abstract notion or idea within the mind, which correlates to a tangible object or referent in the real world. Conversely, the symbol functions as a visual or auditory representation that signifies or stands for the concept associated with the referent. The example provided demonstrates the establishment of a direct relationship between the referent-reference and referent-symbol (**Figure 3**). The first relationship is referred to as referencing, while the second is modelling. The relationship between the symbol and the object is complex as both exist external to the human mind. However, it can be asserted that constructing a building involves translating symbolic design representations (SYMBOL) into physical structures (OBJECT). Human interpretation of the symbols remains crucial until this translation can be carried out entirely by robots. Augmented reality facilitates this interpretation by overlaying the symbols onto real-world imagery. This technology surpasses traditional 2D plans, projections, and virtual reality by eliminating the separation between the symbolic and the real, with the human mind acting as the interface between the two.

It allows stakeholders to visualise and explore proposed designs in their intended physical context, enabling better spatial understanding and design validation. AR can help identify potential conflicts during the design phase, enabling early identification and resolution of issues. In addition, augmented reality can improve on-site construction activities by providing real-time guidance and visual aids, increasing accuracy and efficiency. It also has the potential to revolutionise building management and maintenance by overlaying information about physical components, enabling better asset management and maintenance planning. The prospect of augmented reality in the AEC industry is promising, but several challenges must be overcome for widespread adoption. Technical limitations such as accurate tracking and registration of virtual objects in the real world, interoperability and standardisation of data, cost-effectiveness, and user acceptance are significant hurdles. In addition, privacy and security concerns and a robust infrastructure to support the AR ecosystem are critical.

The chapter explores various user scenarios and applications of augmented reality in building data modelling and highlights the benefits, challenges, and potential solutions for successful implementation. By understanding the opportunities and limitations of augmented reality in the AEC industry, stakeholders can make informed decisions about its adoption, ultimately leading to improved collaboration, increased productivity, and the creation of sustainable and efficient building environments.

2. AEC specifics

The architecture, engineering, and construction industry differs in its products, processes, and teams. Unlike the manufacturing industry, which produces standardised goods, the AEC industry primarily involves constructing unique structures, such as buildings, bridges, and infrastructure projects. These products require customised design and construction approaches to meet specific customer requirements, local codes, and site conditions. Consequently, the AEC industry faces inherent challenges in complexity, coordination, and collaboration:

1. Unique project: Each project has unique requirements, site characteristics, and constraints, resulting in diverse architectural styles, building systems, and materials. This variety poses challenges for materials procurement, construction techniques, and project management, as the industry lacks economies of scale in mass production.

2. Unique processes: The AEC industry encompasses complex processes, from project development to construction and handover. Each construction project follows a unique workflow based on its specific characteristics and requirements. It involves multiple stakeholders, such as architects, engineers, contractors, suppliers, and regulatory agencies, which contribute their expertise throughout the project life cycle. Effective communication, collaboration, and coordination among multidisciplinary teams are essential for successful project delivery.

3. Unique teams: The AEC industry comprises diverse groups of professionals with diverse expertise and responsibilities. Architects prioritise aesthetics, functionality, and space planning, while engineers specialise in structural engineering and technical aspects. Contractors and subcontractors manage the construction process and ensure compliance with regulations. These multidisciplinary teams collaborate to achieve a cohesive project outcome. However, integrating different perspectives, resolving conflicts, and maintaining effective communication can be challenging due to the diverse nature of the groups involved.

The uniqueness of AEC projects, processes, and teams creates a challenging environment with islands of automation, fragmented information, and limited interoperability between stakeholders. Traditional paper-based documentation and manual coordination methods have proven insufficient to manage the complexity of the industry effectively. As a result, the AEC industry has recognised the need for innovative technologies and approaches to address these challenges and improve productivity, efficiency, and collaboration. BIM has emerged as a powerful tool to manage the complexity of unique AEC projects. It enables the creation of a digital representation of the building or infrastructure project, integrating various data and information into a central model. This approach facilitates collaboration, reduces errors, improves coordination, and allows stakeholders to visualise and analyse the project comprehensively and interactively. BIM promotes more efficient information sharing and supports seamless communication and integration between multidisciplinary teams.

The characteristics of the AEC industry, including the production of unique products, the implementation of customised processes, and the involvement of diverse teams, create inherent challenges. The industry has recognised these challenges and embraced innovative technologies, such as BIM, to manage complexity, improve coordination and increase project outcomes. Using advanced tools and methodologies, the AEC industry strives to streamline its processes, foster collaboration, and deliver high-quality, customised projects that meet the ever-evolving needs of clients and society.

3. AR implementation

Implementing AR systems in the AEC sector can be facilitated by adopting an OpenBIM approach. OpenBIM is a collaborative framework for creating and managing BIM data models across software platforms [10]. Integrating AR technology with OpenBIM improves collaboration, data interoperability, and project efficiency. OpenBIM emphasises open standards for data exchange, enabling seamless interoperability between software applications [11]. By adhering to OpenBIM principles, AR platforms can access and use BIM data from multiple sources without compatibility issues. This enhances visualisation and collaboration during design reviews, construction coordination, and facilities management. AR systems based on OpenBIM can be seamlessly integrated into existing BIM workflows. Stakeholders can link AR experiences to specific elements within the BIM model, providing context-specific information.

For example, on-site workers can view instructions or safety guidelines overlaid on physical elements in real-time, increasing productivity and reducing errors. OpenBIM encourages collaboration between disciplines and stakeholders throughout the project lifecycle. Integrating AR into an OpenBIM environment enables shared, immersive experiences. Users can visualise, annotate, and interact with the BIM model in real-time, regardless of location, leading to better coordination and decision-making. OpenBIM and AR systems provide scalability and flexibility. They support integrating multiple software applications and adapting to various hardware platforms and user preferences. This ensures adaptability to evolving technologies and project requirements. Combining AR with OpenBIM allows the capture and documentation of augmented experiences within the BIM model. Users can record AR annotations, instructions, or visualisations linked to specific elements. This facilitates information sharing, handover, and future maintenance and facility management activities.

OpenBIM, based on the Industry Foundation Classes (IFC) standard, can present challenges due to the need for consistent implementation across various BIM software providers. IFC, developed by buildingSMART, is an open file format that facilitates information exchange throughout the building life cycle. It enables seamless communication and data exchange between different software applications, regardless of the platform used. However, IFC has its limitations, including the complexity and size of IFC files, which can impact performance for large-scale projects. Inconsistencies and ambiguities in the standard can also lead to data inconsistencies and inaccuracies when exchanging information between software applications.

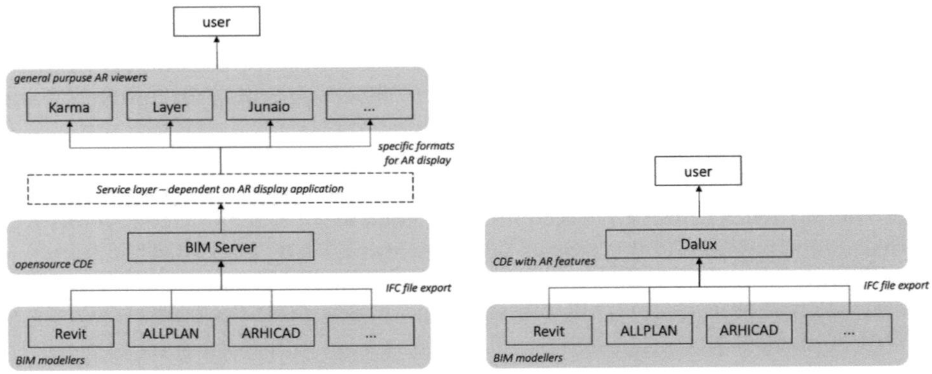

Figure 4.
OpenBIM AR system architecture – (left) using general purpose AR viewers, (right) integrated solution, CDE with AR capabilities.

Many different AR systems follow the above approach: from independent open research systems AR [12] to AR systems that are part of complex shared data environments. Of course, there are clear advantages to using these integrated solutions, but one must also be aware of potential shortcomings, such as locking down data and users (**Figure 4**). The common data environment Dalux [13] was used for this research and testing, as this platform was chosen for the construction project. As with the AR system, there are many options for selecting devices for this purpose. The choice of a device can be based on many additional requirements or factors, such as whether the device is to be used on a construction site, whether it is to be used indoors or outdoors, whether it is a handheld device or a portable device, and whether it is price limited. This research used a handheld device (Samsung Galaxy Table S7 with 5G connectivity). The choice was not optimal, as seen from the description in the next section, but was justified because smart devices (phones, tablets) are accessible, always available, and offer good value for money.

4. Use-case scenarios

As identified by different authors [14, 15], there are three main use-case scenarios for the use of AR within the BIM workflows, including: (1) design – enabling the review of proposed solutions (**Figure 5**), (2) construction – enabling efficient and effective construction progress monitoring [16], and (3) operation – assisting in building maintenance tasks (**Figure 6**) [17]. However, it should be noted that other use-case scenarios warrant consideration [18], for example, optimising layouts, conducting excavations, establishing precise positioning, performing inspections, coordinating tasks, supervising activities, and providing comments. For this research, the focus has been directed towards the construction phase, specifically monitoring construction progress. This use-case investigation aims to explore the potential implementation of AR technology within a selected mobile application for infrastructure projects characterised by extensive distances.

One of the most significant Slovenian infrastructure projects was used for this research. The complete project (managed by 2TDK d.o.o [19], Kolektor Koling d.o.o. [20] as one of the construction companies involved) includes the construction of

Figure 5.
Design phase – (left) visualising a future railway track and (right) a tunnel portal.

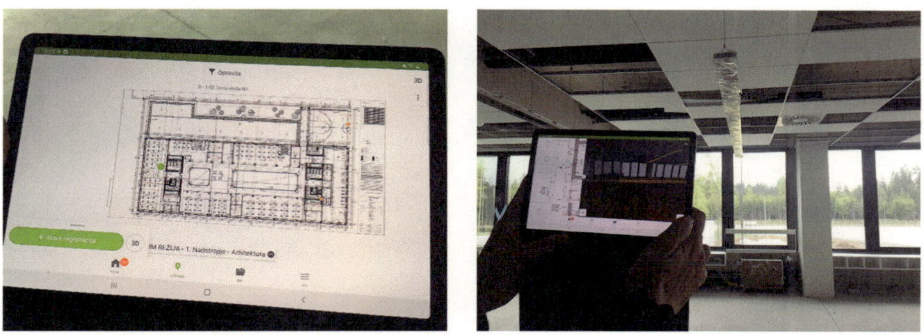

Figure 6.
Operation phase – (left) building maintenance accessing CDE and (right) AR visualisation using a smart device.

more than 27 km of a railway line with eight tunnels (over 20 km in length), two viaducts (424 m and 630 m in length). The same project and construction site were also used by De Hugo Silva et al. [15] to assess the potential of the AR system in the design phase (**Figure** 7).

Figure 7.
One of the largest construction sites of the project.

One of the critical steps in setting up AR visualisation on-site includes positioning and scaling the model according to the natural environment (**Figure 8**). Specific requirements must be met to determine the exact geo-position regarding the BIM model [21]. Unfortunately, this requirement is not always met, representing a significant problem on long construction sites (roads, railways, tunnels).

In order to evaluate the use case, a tablet computer with Dalux mobile application was used at two separate locations, which were a few kilometres apart. The two

Figure 8.
The critical step in setting up AR visualisation – positioning and scaling the BIM model.

Figure 9.
Construction phase – (left) construction site view of retaining wall and railway line, (right) AR view of the planned construction.

locations were georeferenced in the tablet's 2D model, and the relevant information from the 3D model was successfully connected, enabling the visualisation of interactive BIM models directly at the location (**Figure 9**) [22, 23] through the implementation of AR technology. The accompanying images showcase the capability to assess railway installations and track the progress of retaining wall construction according to the planned schedule. Furthermore, AR visualisation proved valuable in comprehending the construction site's overall scale and intricate nature.

Key insights from the analysis include: (1) accurate georeferencing of all BIM models is crucial for their effective utilisation within AR systems, (2) bright environments present challenges when using tablets/phones for AR applications, (3) field orientation and precise positioning and scaling of BIM models can be challenging, (4) AR can provide a valuable tool for gaining an insight of the construction site, and (5) integrated Dalux solution enables the use of AR in practice.

5. Conclusions

The primary focus of this chapter centres on use-case scenarios highlighted by researchers and practitioners in the AEC industry. With the widespread adoption of BIM methodology as the standard [24] for managing construction projects, coupled with the general digitisation of processes and workflows in the AEC industry, there is an opportunity to leverage advanced information and communication technologies, including augmented reality, to enhance AEC workflows characterised by unique products, processes, and teams. It is important to note that the intention is not to assert the superiority or exclusivity of AR over other technologies but rather to highlight its complementary role in managing the lifecycle of a building. An area where AR demonstrates significant potential is in education, as it can generate interest in engineering studies among younger generations. AR facilitates effective communication and decision-making among project stakeholders by enabling instant visualisation. As noted by Wang et al. [25], AR technology inherently facilitates the interaction between real-world and virtual information sources. Within BIM technology, AR allows designers to situate virtual blueprints in a natural environment, provides owners with immersive and interactive experiences, and enables vendors to communicate with different stakeholders effectively.

Table 1 presents the SWOT analysis for integrating augmented reality into BIM workflows, taking into account the AEC specifics discussed earlier, as well as the use cases and key insights from the practical implementation of the technology across various construction sites.

Based on the SWOT analysis, it can be inferred that augmented reality in the AEC sector offers notable strengths in enhanced visualisation, improved collaboration, and on-site support. However, several challenges, such as technical limitations, data interoperability, cost implications, and user acceptance, must be addressed effectively. By leveraging the opportunities presented by augmented reality, such as improved design validation, enhanced client engagement, and streamlined maintenance processes, successful integration within the AEC sector can lead to increased productivity, better project outcomes, and improved operational efficiency.

The AEC industry is currently grappling with a significant scarcity of qualified engineers, which poses obstacles to project delivery and industry growth. The demand for engineering expertise often surpasses the available pool of skilled professionals, resulting in increased workloads, extended project timelines, and

Strengths	**Weaknesses**
• Enhanced visualisation • Gain a comprehensive visual representation • Adjust the object's scale within the application • On-site support and guidance • Improved collaboration and communication	• Challenges in accurately placing objects within the natural environment • Limitations of mobile devices impacting the quality of results • Issues related to transferring large amounts of data • Lack of integration with BIM modellers
Opportunities	**Threats**
• Facilitated knowledge acquisition and transfer • Enhanced validation of designs • Advancements in BIM modelling • Different specialised applications • Real-time remote interaction • Utilisation in maintenance and facility management	• Considerations regarding cost and accessibility • Unavailability of a 3D BIM model • Absence of necessary information • Presence of incorrect information • Factors influencing user acceptance and adoption • Concerns related to privacy and security

Table 1.
SWOT analysis of augmented reality in BIM.

compromised quality. Augmented reality technology has the potential to alleviate this challenge by enhancing the capabilities of existing engineers and empowering less experienced individuals to acquire new skills and knowledge.

Future research endeavours should primarily focus on three key areas. Firstly, there is a need for a better understanding of information transfer processes to determine which tasks could derive the most benefit from AR technology. Additionally, improvements in software solutions are essential to facilitate seamless integration and maximise the potential of augmented reality in the AEC industry. Lastly, there is a requirement to gain a deeper understanding of the information requirements in BIM models. Exploring alternative information flows would be an intriguing avenue for further investigation.

Acknowledgements

The presented research supported the Slovenian Research Agency research program E-construction (E-Gradbeništvo): P2-0210. Their support is gratefully acknowledged.

The author also wishes to thank Metod Gaber, Kolektor d.d. for his support and the 2TDK d.o.o. company for access to the construction sites and for using BIM models in this research.

Conflict of interest

The authors declare no conflict of interest.

DOI: http://dx.doi.org/10.5772/intechopen.1002371

Author details

Matevž Dolenc
Faculty of Civil and Geodetic Engineering, University of Ljubljana, Ljubljana, Slovenia

*Address all correspondence to: mdolenc@fgg.uni-lj.si

References

[1] Eastman C, Teicholz P, Sacks R, Liston K, editors. BIM Handbook: A Guide to Building Information Modeling for Owners, Managers, Designers, Engineers and Contractors. 2nd ed. New Jersey: Wiley; 2011. p. 99. DOI: 10.5130/ajceb.v12i3.2749

[2] Richard D, Chris H. Implementing 'Site BIM': A case study of ICT innovation on a large hospital project. Automation in Construction. 2013;**30**:15-24. DOI: 10.1016/j.autcon.2012.11.024. ISSN 0926- 5805

[3] Patacas J, Dawood N, Mohamad Kassem. BIM for facilities management: A framework and a common data environment using open standards. Automation in Construction. 2020;**120**:103366. DOI: 10.1016/j.autcon.2020.103366

[4] Renzi FV. BIM Application in Construction Management [Thesis]. Barcelona: UPC Universitat Politècnica de Catalunya; 2018

[5] Saar CC, Klufallah M, Kuppusamy S, Yusof A, Shien LC. Mint: Bim integration in augmented reality model. International Journal of Technology. 2019;**10**(7):611-619. DOI: 10.14716/ijtech.v10i7.3278

[6] Wolfartsberger J. Mint: Analysing the potential of virtual reality for engineering design review. Automation in Construction. 2019;**2018**(104):27-37. DOI: 10.1016/j.autcon.2019.03.018

[7] Alizadehsalehi S, Hadavi A, Huang JC. Mint: From BIM to extended reality in AEC industry. Automation in Construction. 2020;**116**(103254):1-13. DOI: 10.1016/j.autcon.2020.103254

[8] Meža S, Turk Ž, Dolenc M. Mint: Measuring the potential of augmented reality in civil engineering. Advances in Engineering Software. 2015;**90**:1-10. DOI: 10.1016/j.advengsoft.2015.06.005

[9] Meža S, Turk Ž, Dolenc M. Integrating ideas, symbols, and physical objects in architecture engineering and construction. In: 11th International Postgraduate Research Conference (IPGRC 2013), 8-10 April 2013, Manchester, UK. Salford: University of Salford, School of the Built Environment; 2013. pp. 297-304

[10] Jiang S, Jiang L, Han Y, Wu Z, Wang N. OpenBIM: An enabling solution for information interoperability. Applied Sciences. 2019;**9**(24):5358. DOI: 10.3390/app9245358

[11] Pan X, Khan AM, Eldin SM, Aslam F, Rehman SKU, Jameel M. BIM adoption in sustainability, energy modelling and implementing using ISO 19650: A review. Ain Shams Engineering Journal. 2023:102252. DOI: 10.1016/j.asej.2023.102252. ISSN 2090-4479

[12] Meža S, Turk Ž, Dolenc M. Component-based engineering of a mobile BIM-based augmented reality system. Automation in Construction. 2014;**42**:1-12. DOI: 10.1016/j.autcon.2014.02.011

[13] DALUX. Available from: https://www.dalux.com/gb/ [Accessed: June 16, 2023]

[14] Shin DH, Dunston PS. Identification of application areas for augmented reality in industrial construction based on technology suitability. Automation in Construction. 2008;**17**(7):882-894. DOI: 10.1016/j.autcon.2008.02.012. ISSN 0926-5805

[15] De Hugo C, Silva A, Gaber M, Dolenc M. Using augmented reality in different BIM workflows. In: Cvetković D, editor. Augmented Reality and Its Application. Rijeka: IntechOpen; 2022. pp. 145-158. DOI: 10.5772/intechopen.99336

[16] Fu M, Liu R. The Application of Virtual Reality and Augmented Reality in Dealing with Project Schedule Risks. 2018. pp. 429-438. DOI: 10.1061/9780784481264.042

[17] Duston PS, Shin H. Key areas and issues for augmented reality applications on construction sites. In: Mixed Reality in Architecture Design and Construction. Netherlands: Springer; 2009. pp. 157-170. DOI: 10.1007/978-1-4020-9088-2_10

[18] Woodward C, Hakkarainen M. Mobile mixed reality system for architectural and construction site visualisation. In: Nee AYC, editor. Augmented Reality – Some Emergent Application Areas. Rijeka: IntechOpen; 2011. pp. 115-130. DOI: 10.5772/26117

[19] 2TDK d.o.o, Available from: https://drugitir.si [Accessed: June 16, 2023]

[20] Kolektor Koling d.o.o, Available from: https://www.kolektorgradbenistvo.si [Accessed: June 16, 2023]

[21] Diakite AA, Zlatanova S. Automatic georeferencing of BIM in GIS environments using building footprints. Computers, Environment and Urban Systems. 2020;**80**:101453. DOI: 10.1016/j.compenvurbsys.2019.101453

[22] Muhammad AA, Yitmen I, Alizadehsalehi S, Celik T. Mint: Adoption of virtual reality (VR) for site layout optimization of construction projects. Teknik Dergi. 2020;**31**(2):9833-9850. DOI: 10.18400/tekderg.423448

[23] Ratajczak J, Riedl M, Matt DT. Mint: BIM-based and AR application combined with location-based management system for the improvement of the construction performance. Buildings. 2019;**9**(118):1-17. DOI: 10.3390/buildings9050118

[24] International Organization for Standardization (ISO). Organization and Digitisation of Information About Buildings and Civil Engineering Works, Including Building Information Modelling (BIM) — Information Management Using Building Information Modelling — Part 1: Concepts and Principles (ISO Standard 19650-1:2018). 2018. Available from: https://www.iso.org/standard/68078.html

[25] Wang J, Wang X, Shou W, Xu B. Mint: Integrating BIM and augmented reality for interactive architectural visualisation. Construction Innovation. 2014;**14**(4):453-476. DOI: 10.1108/CI-03-2014-0019

Chapter 7

Urban Augmented Reality for 3D Geosimulation and Prospective Analysis

Igor Agbossou

Abstract

The advent of augmented reality (AR) has introduced a new era of real-time geosimulation and analysis, particularly in urban planning, spatial design, and architecture. In this chapter, we propose a framework for using urban augmented reality model (UARM) to implement 3D geosimulation and prospective analysis of urban built environments. Our framework leverages advanced technologies, such as computer vision, 3D modeling, and machine learning, to provide a realistic and interactive representation of urban built environments. Using UARM, stakeholders can visualize and analyze the impact of proposed changes to the built environment in real-time. This paper presents the technical specifications and implementation details of our proposed framework and provides case studies demonstrating its effectiveness in urban planning and design. This paper will serve as a guideline for future research in implementation tools for virtual geographic environments (VGE).

Keywords: urban augmented reality, 3D geosimulation, prospective analysis, planning, spatial design

1. Introduction

The field of urban planning, spatial design, and architecture has long relied on various methods and tools to assess the impact of proposed changes in the built environment [1, 2]. However, these traditional approaches often lack interactivity [3, 4] and fail to provide stakeholders with an immersive and realistic understanding of the potential future developments [5, 6]. With the advent of augmented reality (AR) technology, there is a significant opportunity to revolutionize the way we perceive and analyze urban environments [6, 7]. Augmented reality enhances our perception of the physical world by overlaying digital information onto it, thereby providing a seamless integration of virtual and real-world elements. This technology has already made remarkable strides in entertainment and gaming [8–11], but its potential for practical applications in urban planning and land use is yet to be fully explored [6, 7, 9, 11]. In this paper, we propose a framework that leverages urban augmented reality models (UARM) [12] to implement 3D geosimulation and prospective analysis in the context of urban built environments.

The main objective of this framework is to enable stakeholders, including researcher, planners, designers, and decision-makers, to visualize and analyze proposed changes to the built environment in real-time. By combining advanced technologies such as computer vision [13], 3D modeling [9], and machine learning [14, 15], the UARM framework provides a realistic and interactive representation of urban spaces. This allows stakeholders to gain a comprehensive understanding of the potential impacts of their decisions and interventions. One of the key components of the framework is computer vision, which enables the accurate detection and tracking of physical elements in the urban environment. By analyzing live video feeds or stereoscopic images, computer vision algorithms align virtual objects with the real-world context. This ensures a seamless integration of digital information within the physical space, creating an immersive experience for stakeholders. Another important aspect of the UARM framework is the use of high-fidelity 3D models of urban environments. These models capture both the geometric and semantic information of the built environment, providing a realistic representation of buildings, roads, and other urban elements. Techniques such as laser scanning, photogrammetry, and CAD modeling are employed to generate these 3D models using universal scene description (USD) schema and specifications [...].

The main objective of the project was to develop a VGE [16] framework using the augmented reality application development kit (ARKit) [17, 18] for 3D geosimulation and prospective analysis. The research aimed to answer the following questions: 1) How does augmented reality enhance the accuracy and realism of 3D geosimulation in urban areas? 2) What are the effective methods to integrate real-time data into augmented reality for prospective analysis in urban planning and design? 3) How can augmented reality facilitate stakeholder engagement and participatory design in urban planning and design processes? 4) What are the challenges and opportunities of integrating augmented reality with simulation models in the urban environment?

Addressing these questions will contribute to the advancement of augmented reality in 3D geosimulation and prospective analysis of urban areas, leading to improved decision-making, increased stakeholder engagement, and sustainable practices in urban planning and design. After exploring the foundations of augmented reality for urban simulation, including key concepts, enabling technologies, and the limitations of traditional approaches in Section 2, we will clarify and discuss the settings of the experimental application of UARM, focusing on the urban built environment geosimulation area and the use of sensors for data acquisition in Section 3. We present in Section 4, scenario-based prospective analysis results, highlighting the advancements in urban planning through green spaces and urban ecology, as well as urban sustainability through energy-efficient interventions. Section 6 concludes this chapter and future work.

2. Foundations of augmented reality for urban simulation

In recent years, geosimulation based on VGE framework has emerged as a valuable approach in urban planning and design, enabling the simulation and analysis of complex urban systems [19–21]. Geosimulation combines geospatial data, computer modeling, and simulation techniques to replicate and study the dynamics of urban environments. It provides a powerful tool for understanding the complex interactions between various elements of the urban environment, including buildings, transportation systems, and human activities. One of the key applications of geosimulation in urban planning is scenario modeling. By simulating different scenarios, planners and designers can explore the potential impacts of various interventions and policies on the urban landscape. It allows stakeholders to assess factors such as population dynamics, land use changes, transportation patterns, and environmental impacts, providing valuable insights for decision-making. With the advancement of remote sensing technologies and data collection techniques, geospatial data has become more accessible and comprehensive. This wealth of data allows for the creation of realistic 3D models, which serve as the basis for UARM in urban augmented reality applications. The integration of geosimulation and augmented reality offers a unique opportunity to bridge the gap between virtual simulations and the physical world. By overlaying digital information onto the real-world context, augmented reality enhances the understanding and perception of urban simulations. Indeed, AR has emerged as a powerful technology that enhances our perception and interaction with the physical world by overlaying digital information onto real-world environments. In the context of urban simulation, AR offers unique opportunities to visualize, analyze, and interact with urban environments in real-time.

2.1 Definition and key concepts

AR refers to a technology that overlays digital information, such as images, videos, or 3D models, onto the real-world environment, enhancing the user's perception and interaction with their surroundings. Unlike virtual reality, which immerses users in a completely virtual environment, augmented reality supplements the physical world with digital content, creating a blended experience. AR applications typically rely on devices such as smartphones, tablets, smart glasses, or headsets to deliver the augmented experience to users. These devices incorporate cameras, sensors, and displays to capture and augment the real-world environment in real-time. Two fundamental concepts in augmented reality are registration and tracking. Registration involves aligning virtual objects with the real-world context, ensuring they appear in the correct position and orientation. Tracking involves continuously monitoring the user's viewpoint and the physical environment to maintain the spatial consistency of the augmented content.

2.2 Technologies enabling urban augmented reality

Several technologies contribute to the development and implementation of augmented reality systems. Understanding these technologies is crucial for designing effective urban augmented reality solutions. Computer vision [...] plays a vital role in AR by enabling the recognition and understanding of the physical environment. It involves the analysis of visual data [6, 13, 22], such as images or video streams, to detect and track objects, estimate their pose, and extract relevant features [22–24]. Computer vision algorithms facilitate the registration of virtual objects in the real world, allowing for seamless integration and interaction. Accurate and detailed 3D models of the urban environment are essential for realistic and contextually relevant augmentations. Techniques such as laser scanning [12], photogrammetry [12], and CAD modeling enable the creation of high-fidelity 3D models [1, 25]. These models capture the geometric and semantic information of buildings, streets, and other urban elements, forming the foundation for precise and visually consistent augmentations [18, 22, 26]. Machine learning techniques [14, 15, 22, 24] play a crucial role in augmenting reality. They enable object recognition [12, 24], semantic understanding, and real-time tracking. Machine learning algorithms can be trained to recognize and classify urban objects, allowing for intelligent augmentations and predictive analysis [27]. Furthermore, AI-based algorithms can adapt and improve over time, enhancing the accuracy and effectiveness of augmented reality systems. The integration of augmented reality with urban and spatial computing offers new possibilities for analyzing and simulating the built environment. Urban and spatial computing focuses on the interactions between humans, their physical environment, and digital information systems. By combining augmented reality with urban and spatial computing, stakeholders in urban planning and design can visualize and analyze proposed changes to the built environment in real-time. The foundations of augmented reality and its integration with urban and spatial computing offer immense potential for enhancing the field of urban planning, spatial design, and architecture. **Figure 1** illustrates the different components of the process underlying the UARM for prospective analysis.

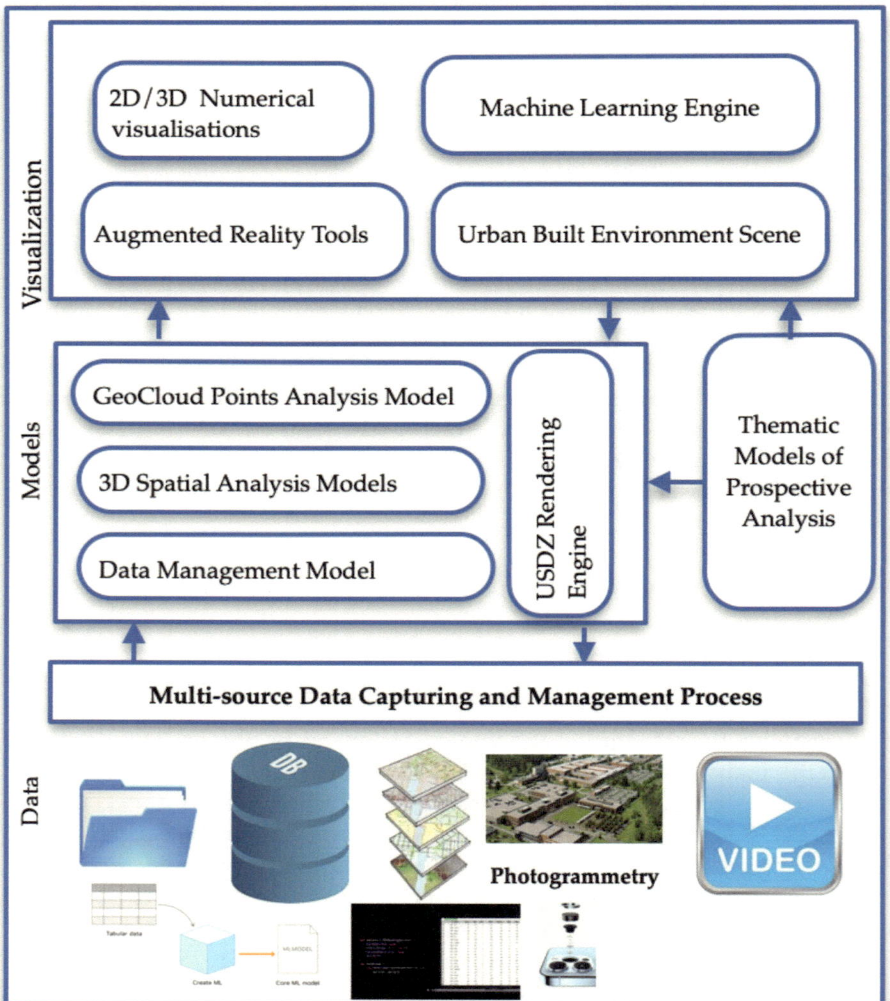

Figure 1.
Urban augmented reality modeling process components.

2.3 Urban prospective analysis and traditional approaches limitations

Urban prospective analysis plays a crucial role in understanding the potential impacts of urban planning and design interventions on the future development of cities. By assessing different scenarios and evaluating their implications, decision-makers can make informed choices to shape sustainable and resilient urban environments. However, traditionalpproaches to prospective analysis in urban planning and design have several limitations that can be overcome by leveraging augmented reality and 3D geosimulation. In **Table 1**, we expose the limitations of traditional approaches and discuss how augmented reality can address these challenges.

3. Experimental application of UARM for prospective analysis

Simulating large-scale urban built environment processes using physically-based rendering in 3D poses significant challenges for modern computing techniques in urban studies and regional planning [28–31]. Urban systems inherently exhibit complexity [3, 19–21, 32], and simulation serves as a tool to comprehend the causes and impacts of events within these systems. Additionally, simulation enables the prediction of future states resulting from specific actions. The level of detail [29, 31, 33, 34] achieved in simulating real system behavior depends on the chosen model. More detailed models with extensive data can provide a more accurate reflection of reality, but their complexity directly affects computational time required for model changes.

3.1 Urban built environment geosimulation area

In this experimental study, our URAM framework was applied to a section of a newly constructed housing estate in Belfort, France (**Figure 2**). The development project comprises 25 plots ranging from 600 to 900 m2 for individual houses. It's called "Jardins du MONT," and depicts a contemporary with high-quality architectural design, conveniently located within a 10-minute travel distance from the city center of Belfort via car, bus, or bike. It is also situated within a short walking distance from the bustling "Techn'Hom" business park, which houses major companies such as GE and Alstom. The area offers a serene and green urban environment, providing exceptional views of Belfort and its fortifications.

The research work undertaken in this study focuses on 3D spatial analysis, the temporal evolution of new housing estates, and the implementation of smart city concepts using scientific tools in artificial intelligence. Considering the ongoing development of this specific urban area, it was deemed appropriate to apply the URAM to conduct a prospective analysis of the urban built environment.

3.2 Sensor for data acquisition

In the context of data acquisition for urban augmented reality modeling, the choice of sensor plays a crucial role in capturing accurate and high-quality data. With the advancement of technology, a wide range of sensors are available for collecting geospatial data in urban environments. These sensors enable researchers to capture various types of data, including spatial coordinates, 3D point clouds, images, and depth information. One of the widely used sensors for data acquisition is the Light Detection and Ranging (LiDAR) scanner. LiDAR scanners emit laser pulses

Traditional approaches limitations		Role of AR in addressing limitations	
Lack of Visual Realism	Traditional prospective analysis methods often rely on 2D maps, diagrams, or static renderings to visualize future scenarios. This limited visual representation can make it challenging for stakeholders to fully comprehend and evaluate the proposed interventions. It may lead to misunderstandings or overlooking critical aspects of the design, hindering effective decision-making.	Enhanced Visual Realism	AR can overcome the lack of visual realism in traditional prospective analysis methods. By overlaying virtual content onto the physical environment, augmented reality provides a more immersive and realistic representation of future scenarios. Stakeholders can visualize proposed interventions as if they already exist in the real world, enabling them to better understand and evaluate the potential outcomes.
Difficulty in Spatial Contextualization	Traditional prospective analysis methods often struggle to provide a comprehensive spatial contextualization of proposed interventions. 2D representations fail to capture the three-dimensional nature of urban environments, making it challenging to assess the impact of interventions on the existing built environment, transportation networks, and open spaces. This limitation restricts the ability to evaluate design alternatives in their proper spatial context.	Spatial Contextualization	Coupled with 3D geosimulation, AR enables the integration of proposed interventions into the existing urban fabric. Stakeholders can experience the design in its proper spatial context, observing how it interacts with surrounding buildings, infrastructure, and natural elements. This spatial contextualization facilitates a more comprehensive understanding of the design's impact on the urban environment.
Limited Stakeholder Engagement	Traditional approaches to prospective analysis often lack effective stakeholder engagement. Decision-makers and stakeholders are typically presented with finalized designs or scenarios, leaving little room for meaningful participation and input. This limited engagement can lead to a lack of ownership, decreased satisfaction, and potential conflicts among stakeholders.	Improved Stakeholder Engagement	AR facilitates enhanced stakeholder engagement in prospective analysis. By providing an interactive platform, augmented reality allows stakeholders to actively participate in the design process. They can explore and manipulate virtual objects, test different scenarios, and provide real-time feedback. This participatory approach fosters collaboration, empowers stakeholders, and promotes a sense of ownership in decision-making.

Traditional approaches limitations		**Role of AR in addressing limitations**	
Time-Intensive Iterative Process	Traditional prospective analysis methods tend to have a lengthy and resource-intensive iterative process. As stakeholders provide feedback and propose modifications, multiple iterations of analysis and redesign are required, leading to extended project timelines and increased costs. This inefficiency can impede the agility and responsiveness needed in dynamic urban planning processes.	Streamlined Iterative Process	AR can streamline the iterative process of prospective analysis. Stakeholders can make design modifications and instantly visualize their impact, reducing the time and effort required for multiple iterations. Augmented reality also supports rapid prototyping and scenario testing, enabling decision-makers to evaluate design alternatives efficiently and make timely adjustments.

Table 1.
Traditional approaches limitations and the role of AR in addressing them.

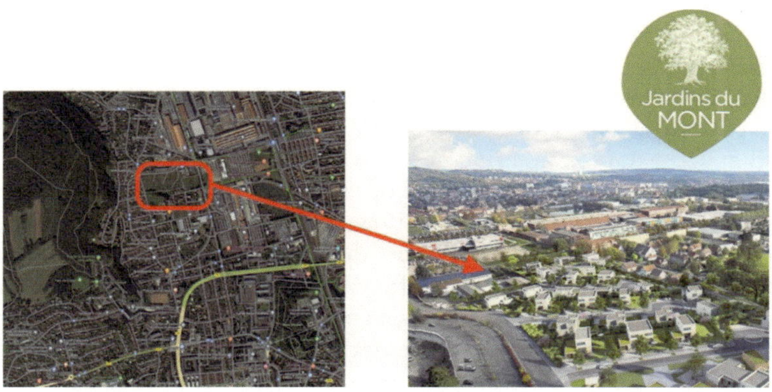

Figure 2.
Experimental study area "Jardins du MONT", Belfort (France).

and measure the time it takes for the laser to return after hitting objects in the environment. This data is then used to generate precise 3D point clouds, which are essential for creating detailed urban models. Another commonly employed sensor is the Global Navigation Satellite System (GNSS) receiver, which uses satellite signals to determine accurate spatial coordinates. GNSS receivers provide location information with high precision and are often used in conjunction with other sensors to enhance data acquisition accuracy. Mobile sensors such as smartphones and tablets have also gained popularity in recent years [35, 36]. These devices are equipped with advanced cameras and sensors, including RGB cameras and depth sensors. The cameras capture high-resolution images, while the depth sensors provide distance measurements from the sensor to objects in the scene. The combination of these sensors enables researchers to capture both visual and depth data for urban modeling. Furthermore, aerial platforms such as drones and aircraft equipped with sensors are utilized for

Figure 3.
iPhone 13 Pro Max used as sensor for data acquisition.

data acquisition in larger-scale urban areas. These platforms enable the collection of geospatial data from a bird's-eye view, providing a broader perspective of the urban environment. Sensors such as LiDAR scanners and RGB cameras mounted on drones or aircraft allow for efficient data capture over large areas. The choice of sensor depends on various factors, including the specific data requirements, the scale of the study area, budget constraints, and logistical considerations. Researchers must carefully evaluate the capabilities and limitations of different sensors to ensure the acquisition of accurate and comprehensive data for urban augmented reality modeling. For our specific data acquisition needs, we selected the iPhone 13 Pro Max as our sensor of choice. This smartphone model offers a range of advanced features that contribute to the quality of the captured images. The sensors integrated into modern smartphones provide capabilities that meet the requirements for data acquisition in photogrammetry, making them suitable for capturing high-quality images for 3D modeling [37]. These features include wide color capture for photos and live photos, lens correction to ensure accurate representations, retina flash for enhanced lighting conditions, auto image stabilization for reducing blurriness, and burst mode for capturing multiple frames in quick succession. The combination of these features makes the iPhone 13 Pro Max well-suited for our research purposes. **Figure 3** visually depicts the iPhone 13 Pro Max as the primary sensor utilized in our experimental study, highlighting its role in capturing the necessary data for our urban augmented reality modeling efforts.

3.3 Enhanced data collection for UARM approach

When capturing images for augmented reality, a specific region of the image sensor is utilized, specifically an area of 3840x2880 pixels on the iPhone 13 Pro. To optimize image processing and memory usage, a technique called binning is applied [38, 39]. Binning involves averaging the pixel values within a 2x2 pixel region and replacing them with a single pixel. This approach offers two significant benefits. Firstly, it reduces the image dimensions by a factor of two, resulting in downscaled images of 1920x1440 pixels. This reduction in size allows for efficient memory consumption and processing power, enabling the camera to operate at up to 60 frames per second while freeing up resources for rendering. Secondly, binning mitigates the impact of sensor noise, making it advantageous in low-light environments. To project the captured images from the 2D image plane into the 3D world, geometric distortion caused by lens imperfections must be corrected. Lens distortion is modeled using a one-dimensional lookup table, consisting of evenly distributed 32-bit float values along a radius from the distortion center to a corner. Each value represents a magnification factor applied to the radius, assuming symmetrical lens distortion [40].

To generate detailed 3D models from real-world photographs using computer vision technology, photos of the urban built environment are taken from various angles using an iPhone. Multiple images are captured, ensuring sufficient overlap for accurate landmark matching and successful 3D reconstruction. Sequential images are positioned to have a 70% overlap or more ($0.7 \leq$ overlap ≤ 0.9) to ensure robust reconstruction [33, 41] as shown in **Figure 4**. Insufficient overlap can lead to reconstruction failures or low-quality augmented reality models. Maintaining a narrow aperture setting to achieve crisp focus is recommended [42, 43]. The spatial precision between image pairs and the density of chromatic textures significantly contribute to the quality of the collected images for 3D reconstruction of urban environments. Key factors that ensure high-quality input data [33, 41–44] are summarized in **Table 2**.

For this experimental study, a photographic database consisting of 800 photos captured in compliance with the overlap constraints was created to feed the model. The database comprises 799 calibrated image pairs, which are sorted based on the constraints of stereovision image matching. **Figure 5** illustrates a sample of the captured data, indicating the reading direction of the photos from start to end. The number of pictures required for an accurate 3D representation varies depending on the quality of the image pairs, the complexity and size of the built environment. It is crucial to adhere to the recommended overlap and aperture settings to ensure the generation of high-quality augmented reality models [33, 41–43]. All the urban built

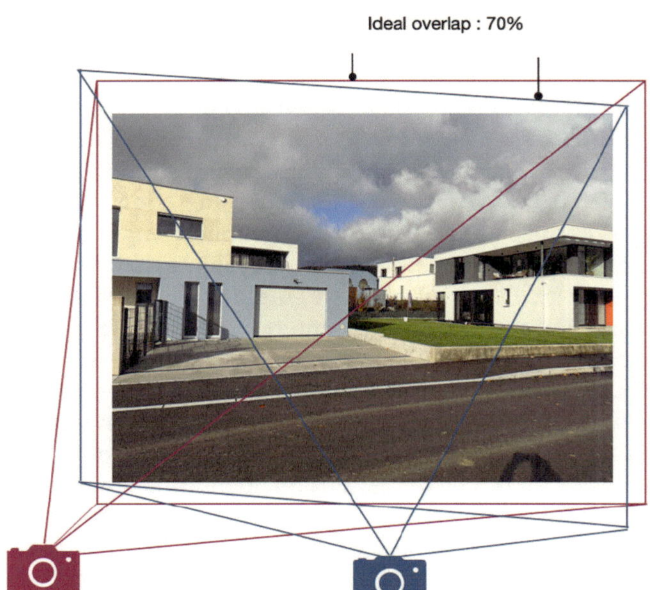

Figure 4.
Ideal overlap to respect when capturing urban built environment.

Factor	Description	Fuzzy threshold value
Range or depth	Distance between camera and scene	Low
Sensor quality	The resolution of de sensor	High
Overlap	Superposition rate between two consecutive photographs	$0.7 \leq$ overlap ≤ 0.9
Image texture	Texture and texture variance	High

Table 2.
Key factors affecting photogrammetric input images quality for URAM.

environment visual features were rendered using USD standard and specifications [6]. The steps and workflow needed to create USD files describing urban 3D visual features are summarized in **Table 3**.

4. Scenario based prospective analysis results

One of the core functionalities of the UARM is the ability to perform 3D geo-simulation and prospective analysis. The framework incorporates computational models and simulation algorithms to simulate the behavior and dynamics of urban systems (**Figure 6**). These models can include factors such as population growth, transportation flows, land use patterns, and environmental factors. By running simulations based on different design scenarios, the UARM enables decision-makers to evaluate the potential impacts of proposed interventions and make informed decisions. Analytics capabilities are also integrated into the UARM, allowing for quantitative and qualitative analysis of the simulation results such as, transportation Infrastructure Expansion, high-density mixed-use development, green spaces and urban ecology, energy efficiency measures. In this chapter we focus on the augmented reality simulation results related to green spaces and urban ecology and energy efficiency.

4.1 Advancing urban planning through green spaces and urban ecology

In contemporary urban planning and design, the integration of green spaces and the promotion of urban ecology are paramount considerations [...]. This prospective scenario explores the potential benefits and implications of incorporating green spaces within the urban environment, aiming to evaluate their impact on various aspects of urban ecology. To achieve this, we employ the UARM approach to create a virtual representation of the urban area of interest. By integrating accurate 3D models of existing structures with virtual green spaces, we can visualize and assess their potential contributions to urban ecology. The envisioned green spaces encompass a range of elements, including parks, community gardens, urban forests, green roofs, and vertical gardens. Through the visualization and assessment provided by UARM, we can analyze the impact of these green spaces on multiple dimensions of urban ecology. For instance, by considering vegetation types, tree canopies, and pollutant dispersion models, we can estimate the potential reduction in air pollution levels. This information facilitates an understanding of how green spaces can contribute to mitigating air pollution and creating healthier urban environments for residents. Furthermore, by incorporating virtual flora and fauna, stakeholders can observe the potential habitats created by the green spaces and evaluate their suitability for

DOI: http://dx.doi.org/10.5772/intechopen.1002352

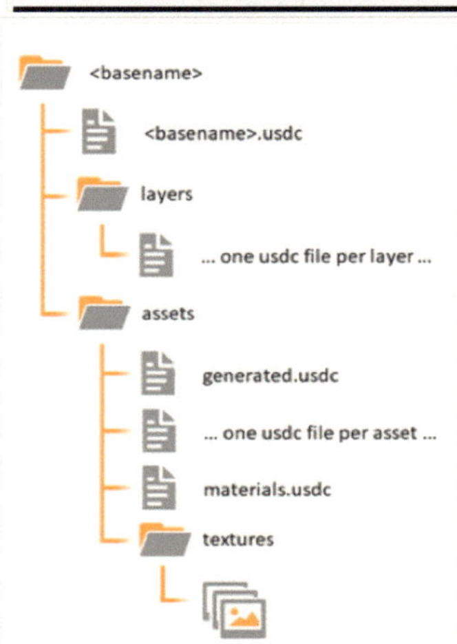


```
#usda 1.0

def Xform "City" {
   def Mesh "Buildings" {
      int[] faceVertexCounts = [4, 4, 4, 4]
      int[] faceVertexIndices = [0, 1, 2, 3, 4, 5, 6, 7, 8, 9, 10, 11, 12, 13, 14, 15]
      float3[] points = [(0, 0, 0), (10, 0, 0), (10, 10, 0), (0, 10, 0),
                         (0, 0, 10), (10, 0, 10), (10, 10, 10), (0, 10, 10),
                         (0, 0, 20), (10, 0, 20), (10, 10, 20), (0, 10, 20),
                         (0, 0, 30), (10, 0, 30), (10, 10, 30), (0, 10, 30)]
      int[] featureIDs = [0, 1, 2, 3]  # Feature IDs for each building
      float[] featureValues = [0.8, 0.6, 0.4, 0.9]  # Feature values for each building
   }

   def Points "Trees" {
      float3[] points = [(5, 0, 0), (0, 0, 5), (5, 0, 10), (10, 0, 5)]
      int[] featureIDs = [0, 1, 2, 3]  # Feature IDs for each tree
      float[] featureValues = [0.2, 0.5, 0.3, 0.7]  # Feature values for each tree
   }

   def Mesh "Roads" {
      int[] faceVertexCounts = [2, 2, 2]
      int[] faceVertexIndices = [0, 1, 2, 3, 4, 5]
      float3[] points = [(0, 0, 0), (10, 0, 0), (0, 0, 10), (10, 0, 10), (0, 0, 20), (10, 0, 20)]
      int[] featureIDs = [0, 0, 1]  # Feature IDs for each road segment
      float[] featureValues = [0.8, 0.6, 0.4]  # Feature values for each road segment
   }
}
```

Figure 5.
Dataset sample for URAM with the USD schema files.

supporting diverse species. This scenario provides insights into the potential increase in biodiversity and ecological connectivity within the urban context. Through simulations that simulate the introduction of green roofs, vertical gardens, and shaded

Step and workflow	Description
Photogrammetry Processing	To process the set of 800 overlapped photographs and generate a 3D point cloud or mesh representing the urban environment, we used Apple ARKit in conjunction with Reality composer Swift programming language.
Geometry Conversion	Convert the 3D point cloud and mesh into a suitable format compatible with USD, such as .usdz and .usda.
USD Scene Assembly	Creation of new USD files (.usda) using a text Apple Reality converter and define the initial stage and layer structure of the scene and import the geometry as a reference or as a direct asset.
Visual Feature Modeling	Within the USD file, one defines the visual features of the urban environment using USD's schema and attribute system. This includes specifying materials, textures, shading parameters, and any other visual properties.
Hierarchy and Organization	Arrange the visual features in a hierarchical structure that reflects the urban environment's spatial relationships. This involve grouping buildings, roads, vegetation, and other elements into separate layers or sublayers.
Metadata Annotation	Enhance the USD files with metadata annotations to capture additional information about the urban features. This includes attributes like building heights, material properties, semantic labels, or any other relevant data.

Table 3.
Steps and workflow needed to create USD files describing urban 3D scene.

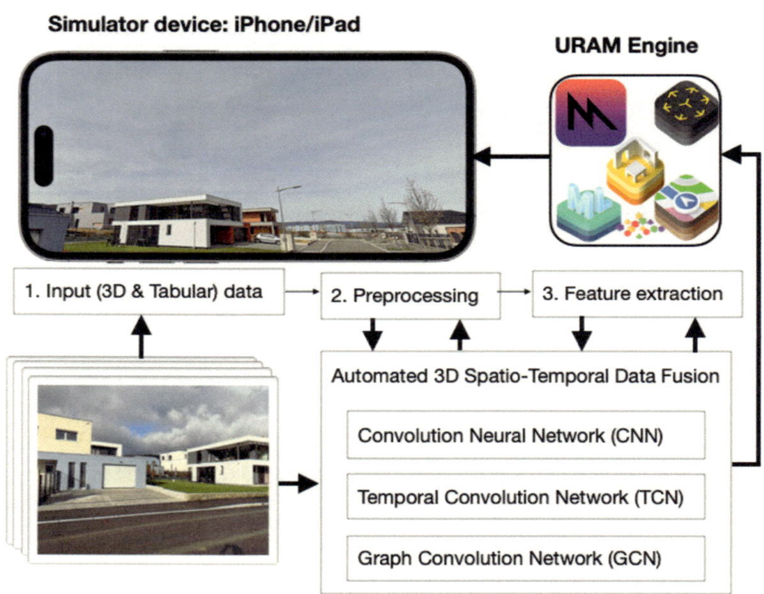

Figure 6.
Scenario based prospective analysis components.

areas, stakeholders can observe the potential reduction in surface temperatures and intensity of the urban heat island effect. Such analyses aid in identifying strategies to mitigate heat-related issues and enhance the thermal comfort of urban residents.

The Green Spaces and Urban Ecology scenario, enabled by UARM, serves as a valuable tool for urban planners and designers. It allows them to visualize and assess the potential benefits of incorporating green spaces within the urban environment, promoting evidence-based decision-making, and fostering stakeholder engagement. By leveraging this approach, we can facilitate the creation of sustainable and livable cities, where the integration of green spaces enhances the overall quality of urban life.

4.2 Advancing urban sustainability through energy-efficient interventions

The implementation of energy-efficient measures within the urban environment has the potential to create a significant impact [1, 4, 9]. This scenario aims to assess the reduction in energy consumption, environmental impact, and economic viability associated with these interventions. By leveraging the capabilities UARM, we visualize the virtual representation of the urban area and explore the potential changes resulting from energy efficiency measures.

Through virtual overlays, we illustrate retrofitted buildings, solar panels, wind turbines, and electric vehicle charging stations, among other elements. The augmented reality environment allows stakeholders to interact with these virtual elements and evaluate their impact on energy usage, carbon emissions, and cost savings. By integrating real-time energy data and building energy models, we can observe the potential reduction in energy consumption resulting from different interventions. This analysis aids in identifying areas with high energy demand and evaluating the effectiveness of proposed energy-saving strategies.

Furthermore, by incorporating cost data, energy pricing models, and return on investment calculations, we assess the financial implications of implementing various interventions. This analysis enables the prioritization of energy-saving measures that provide the greatest economic benefits and cost-effectiveness for urban development projects. By visualizing energy-saving measures in the augmented reality environment, stakeholders gain a better understanding of the associated benefits and actively participate in the decision-making process.

The augmented experience offered by UARM facilitates meaningful discussions, raises awareness, and promotes the adoption of sustainable behaviors among residents, businesses, and communities. By visualizing the potential outcomes of energy-efficient interventions, stakeholders are empowered to make informed choices and actively contribute to urban sustainability.

5. Conclusion and future directions

In this chapter, we have presented the application of Urban Augmented Reality Model (UARM) for 3D geosimulation and prospective analysis in urban planning and design. We have explored the foundations of augmented reality for urban simulation, discussed the technologies enabling UARM, and highlighted the limitations of traditional approaches in urban prospective analysis. Furthermore, we have presented the experimental application of UARM, focusing on the urban built environment geosimulation area, sensor utilization for data acquisition, and enhanced data collection for UARM approach.

The scenario-based prospective analysis results have demonstrated the effectiveness of UARM in advancing urban planning through the integration of green spaces and urban ecology, as well as promoting urban sustainability through energy-efficient interventions. These findings provide valuable insights for decision-makers and urban designers in understanding the potential impacts of proposed changes in the built environment and making informed choices for sustainable urban development.

There are several avenues for future research and development in the field of urban augmented reality and geosimulation. Firstly, further advancements in sensor technologies and data collection techniques can enhance the accuracy and realism of UARM models. This can include the integration of more comprehensive environmental data, real-time monitoring systems, and advanced sensing technologies for capturing finer details of the urban environment. Additionally, the incorporated machine learning and artificial intelligence algorithms can enhance the predictive capabilities of UARM, allowing for more accurate and reliable analysis of prospective scenarios. This can enable stakeholders to anticipate the long-term impacts of urban interventions, optimize resource allocation, and facilitate data-driven decision-making. Moreover, exploring the scalability of UARM to larger urban areas and complex urban systems is an important direction for future research. This includes addressing computational challenges, developing efficient algorithms for handling large-scale geospatial data, and exploring distributed computing approaches for real-time geosimulation and analysis.

Our future research and development efforts focus on further refining the UARM framework for addressing scalability challenges and enhance its capabilities in supporting advanced urban visual analytics [45–47].

Author details

Igor Agbossou
Université de Franche-Comté, Laboratoire ThéMA UMR 6049, IUT NFC, France

*Address all correspondence to: igor.agbossou@univ-fcomte.fr

References

[1] Biljecki F, Ledoux L, Stoter J, Vosselman G. The variants of an LOD of a 3D building model and their influence on spatial analyses. ISPRS Journal of Photogrammetry and Remote Sensing. 2016;**116**:42-54. DOI: 10.1016/j.isprsjprs.2016.03.003

[2] Li L, Tang L, Zhu H, Zhang H, Yang F, Qin W. Semantic 3D modeling based on CityGML for ancient Chinese- style architectural roofs of digital heritage. ISPRS International Journal of Geo-Informatics. 2017;**6**(5):132. DOI: 10.3390/ijgi6050132

[3] Benenson I, Torrens P. Geosimulation: Automata-based Modeling of Urban Phenomena. Chichester: Wiley; 2002

[4] Ledoux H. val3dity: Validation of 3D GIS primitives according to the international standards. Open Geospatial Data, Software Stand. 2018;**3**:1. DOI: 10.1186/s40965-018-0043-x

[5] Sinyabe E, Kamla V, Tchappi I, Najjar Y, Galland S. Shapefile-based multi-agent geosimulation and visualization of building evacuation scenario. Procedia Computer Science. 2023;**220**:519-526. DOI: 10.1016/j.procs.2023.03.066

[6] USDZ. Interopérabilité 3D autour du format de Réalité Augmentée. Available online: https://www.cadinterop.com/fr/les-formats/maillage/usdz.html# [Accessed: June 2, 2023]

[7] OGC CityGML 3.0 Conceptual Model. Available online: https://github.com/opengeospatial/CityGML-3.0CM [Accessed: Mai 7, 2022]

[8] Jung J, Hong S, Yoon S, Kim J, Heo J. Automated 3D wireframe modeling of indoor structures from point clouds using constrained least-squares adjustment for as-built BIM. Journal of Computing in Civil Engineering. 2016;**30**(4):2016. DOI: 10.1061/(ASCE)CP.1943-5487.0000556

[9] Bonczak B, Kontokosta CE. Large-scale parameterization of 3D building morphology in complex urban landscapes using aerial LiDAR and city administrative data. Computers, Environment and Urban Systems. 2019;**73**:126-142. DOI: 10.1016/j.compenvurbsys.2018.09.004

[10] Bielefeldt BR, Reich GW, Beran PS, Hartl DJ. Development and validation of a genetic L-System programming framework for topology optimization of multifunctional structures. Computers & Structures. 2019;**218**:152-169. DOI: 10.1016/j.compstruc.2019.02.005

[11] Henderson P, Ferrari V. Learning single-image 3D reconstruction by generative modeling of shape, pose and shading. International Journal of Computer Vision. 2020;**128**:835-854. DOI: 10.1007/s11263-019-01219-8

[12] Agbossou I. Fuzzy photogrammetric algorithm for city built environment capturing into urban augmented reality model. Artificial Intelligence. 2023. DOI: 10.5772/intechopen.110551

[13] Weinmann M. Visual features – From early concepts to modern computer vision. In: Farinella G, Battiato S, Cipolla R, editors. Advanced Topics in Computer Vision. London: Advances in Computer Vision and Pattern Recognition. London: Springer; 2013. DOI: 10.1007/978-1-4471-5520-1_1

[14] Song C, Lin Y, Guo S, Wan H. Spatial–temporal synchronous graph

convolutional networks: A new framework for spatial–temporal network data forecasting. Proceedings of the AAAI Conference on Artificial Intelligence. 2020;**34**(01):914-921. DOI: 10.1609/aaai.v34i01.5438

[15] Guangyin J, Qi W, Cunchao Z, Yanghe F, Jincai H, Xingchen H. Urban fire situation forecasting: Deep sequence learning with spatio-temporal dynamics. Applied Soft Computing. 2020;**97**(Part B):106730. DOI: 10.1016/j.asoc.2020.106730

[16] You L, Lin H. A conceptual framework for virtual geographic environments Knowledge engineering. International Architect Photogramming and Remote Sense Spatial Information Science. 2016;**XLI-B2**:357-360. DOI: 10.5194/isprs-archives-XLI-B2-357-2016

[17] AppleARKit. More to explore with ARKit. AppleARKit. 2017. Available at: https://developer.apple.com/documentation/arkit [Accessed: June 11, 2023]

[18] Wang ZB, Ong SK, Nee AYC. Augmented reality aided interactive manual assembly design. International Journal of Advanced Manufacturing Technology. 2013;**69**:1311-1321. DOI: 10.1007/s00170-013-5091-x

[19] Batty M. Cities and Complexity. Cambridge: MIT Press; 2005

[20] Batty M, Torrens P. Modeling and prediction in a complex world. Futures. 2005;**37**:745-766

[21] Portugali J. Self-organization and the City. New York: Springer-Verlag; 2000

[22] Huang MQ, Ninić J, Zhang QB. BIM, machine learning and computer vision techniques in underground construction: Current status and future perspectives'. Tunneling and Underground Space Technology. 2021;**2021**:108. DOI: 10.1016/j.tust.2020.103677

[23] Zheng Y, Capra L, Wolfson O, Yang H. Urban computing: Concepts, methodologies, and applications. ACM Transactions on Intellectual System and Technology. 2014;**5**:3. DOI: 10.1145/2629592

[24] Rao J, Qiao Y, Ren F, Wang J, Du Q. A mobile outdoor augmented reality method combining deep learning object detection and spatial relationships for geovisualization. Sensors. 2017;**17**(9):1951. DOI: 10.3390/s17091951

[25] Liao T. Standards and their (recurring) stories: How augmented reality markup language was built on stories of past standards. Science, Technology, & Human Values. 2020;**45**(4):712-737. DOI: 10.1177/0162243919867417

[26] Claudia M, Jung T. A theoretical model of mobile augmented reality acceptance in urban heritage tourism. Current Issues in Tourism. 2018;**21**(2):154-174. DOI: 10.1080/13683500.2015.1070801

[27] Gautier J, Brédif M, Christophe S. Co-visualization of air temperature and urban data for visual exploration. In: 2020 IEEE Visualization Conference (VIS). Salt Lake City, UT, USA; 2020. pp. 71-75. DOI: 10.1109/VIS47514.2020.00021

[28] Liliana B, Luca C, Franco C, Giuseppe R, editors. Future Cities and Regions. Simulation, Scenario and Visioning, Governance and Scales. New York, Heidelberg: Springer; 2011

[29] Verma JK, Paul S, editors. Advances in Augmented Reality and Virtual

Reality. Singapore: Springer; 2022. p. 312. DOI: 10.1007/978-981-16-7220-0

[30] Gustavo A et al. Procedural modeling applied to the 3D city model of bogota: A case study. Virtual Reality & Intelligent Hardware. 2021;**3**(5):423-433. DOI: 10.1016/j.vrih.2021.06.002

[31] Peeters A, Etzion Y. Automated recognition of urban objects for morphological urban analysis. Computers, Environment and Urban Systems. 2012;**36**(6):573-582

[32] Berrou JL, Beecham J, Quaglia P, Kagarlis MA, Gerodimos A. Calibration and validation of the Legion simulation model using empirical data. In: Waldau N, Gattermann P, Knoflacher H, Schreckenberg M, editors. Pedestrian and Evacuation Dynamics. New York: Springer Verlag; 2007. pp. 155-156

[33] Anders K-H. Level of detail generation of 3D building groups by aggregation and typification. In: International Cartographic Conference. Vol. 2. 2005. p. 32

[34] Johannes E et al. Procedural modeling of architecture with round geometry. Computers & Graphics. 2017;**64**:14-25. DOI: 10.1016/j.cag.2017.01.004

[35] Biljecki F, Ledoux H, Stoter J. Generating 3D city models without elevation data. Computers, Environment and Urban Systems. 2017;**64**:1-18

[36] Gnana OV, Karthikeyan SK, Padmanaban S, editors. Smart Buildings Digitalization. Case Studies on Data Centers and Automation. Boca Raton: CRC Press; 2022. p. 314. DOI: 10.1201/9781003240853

[37] Cherdo L. The 8 Best 3D Scanning Apps for Smartphones and IPads in 2019. 2019. Available from: https://www.aniwaa.com/buyers-guide/3d-scanners/best-3d-scanning-apps-smartphones/ [Accessed: May 12, 2022]

[38] Liu Y, Wang W, Xu X, Guo X, Gong G, Lu H. Lightweight real-time stereo matching algorithm for AI chips. Computer Communications. 2022. DOI: 10.1016/j.comcom.2022.06.018

[39] Yuan W, Meng C, Tong X, Li Z. Efficient local stereo matching algorithm based on fast gradient domain guided image filtering. Signal Processing: Image Communication. 2021;**95**:116280. DOI: 10.1016/j.image.2021.116280

[40] Liu Y, Wang W, Xu X, Guo X, Gong G, Lu H. Lightweight real-time stereo matching algorithm for AI chips. Computer Communications. 2023;**199**:210-217. DOI: 10.1016/j.comcom.2022.06.018

[41] Kim T-H et al. Smart city and IoT. Future Generation Computer Systems. 2017;**76**:159-162. DOI: 10.1016/j.future.2017.03.034

[42] Yonghuai L et al. 3D Imaging, Analysis and Applications. Second ed. Switzerland: Springer; 2022. DOI: 10.1007/978-3-030-44,070-1

[43] Xiang W et al. A novel reversible image data hiding scheme based on pixel value ordering and dynamic pixel block partition. Information Sciences, Volume. 2015;**310**(2015):16-35. DOI: 10.1016/j.ins.2015.03.022

[44] Wilm, J., Aanæs, H., Larsen, R., & Paulsen, R. R, Real Time Structured Light and Applications. Kgs. Lyngby: Technical University of Denmark (DTU), 2016 (DTU Compute PHD-2015; No. 400). Available from: https://core.ac.uk/

download/pdf/43255252.pdf [Accessed: October 16, 2021]

[45] A Tool for Exploring Urban Visual Analytics Studies. Available at: https://urban-va-survey.github.io/ [Accessed: June 7, 2022]

[46] Li C, Baciu G, Wang Y, Chen J, Wang C. DDLVis: Real-time visual query of spatiotemporal data distribution via density dictionary learning. IEEE Transactions on Visualization & Computer Graphics. 2022, 2022;**28**(01):1062-1072. DOI: 10.1109/TVCG.2021.3114762

[47] Pumain D, Sanders L, Saint-Julien T. Villes et auto-organisation. Paris: Economica; 1989

Chapter 8

Cloud Management of Pumping Systems Using Digital Twins Supported by Augmented Reality

Juan David Rueda Martínez, Gonzalo Andrés Moreno Olano, Sebastian Roa-Prada, Hernán Darío Mantilla Hernández and Nestor Fernando Saavedra Trujillo

Abstract

Smart pumping systems of tomorrow will feature pumps and drives that respond to real-time changes downstream to keep operations at high efficiency and meet growing performance demand. A key component of a smart pumping system is its digital twin, an exact 3D digital copy of the facility. A digital twin enhanced with Augmented Reality (AR) encompasses as-built facility data captured with 3D scanning devices, as well as precise measurement data collected on the actual rigs with high precision instruments. An interactive model based on augmented reality allows the autonomous and efficient use of pumping systems. It provides clear instructions for the step-by-step management of the system. In addition, it shows relevant information with the exploded views of the components for a better understanding of the operation of the equipment. This research is about the interconnection of the digital twin of pumping systems with the real-world using automation and augmented reality systems. In this project, a local area network is configured to exert control and monitoring on an industrial PLC. This PLC controls a test bench with two centrifugal pumps by means of a web page. An augmented reality application is also developed in Unity 3D with the Vuphoria SDK integration.

Keywords: augmented reality, Unity, digital twin, pumping systems, automation, software development

1. Introduction

Augmented reality (AR) is a technology that has attracted the attention of both researchers and industry professionals. With its ability to combine virtual elements with the real world, AR has opened lots of possibilities across various sectors, transforming the way we interact with information, objects, spaces, and people. Some of the applications of augmented reality in different fields are:

In the field of education, AR has brought new methods to conventional learning. Students now can visualize complex concepts, historical events, or scientific processes through virtual overlays. By augmenting their physical surroundings, AR creates an immersive and engaging learning experience. Moreover, AR facilitates remote collaboration, allowing students and teachers to interact in real-time, regardless of their physical location [1].

From medical training to patient care and surgery, augmented reality has revolutionized medical services in several aspects. Surgeons can now access vital patient information and visualize medical imaging data, which can be overlaid onto the patient's body during surgical procedures [2]. This not only enhances precision but also reduces risks. Medical students also benefit from AR, as they can practice complex surgeries in a virtual environment, gaining invaluable practical experience before operating on real patients. AR-based applications have also found their way into rehabilitation, providing patients with interactive exercises and visual feedback to aid in their recovery.

In the commercial sector, AR has been used to enable the end user to visualize products before purchasing them. Virtual try-on applications, for example, allow shoppers to see how clothing, accessories, or even furniture would look on them or in their homes [3, 4].

Retailers and politicians have also begun to leverage AR for interactive marketing campaigns, engaging customers with immersive brand experiences [5].

Additionally, AR has found utility in navigation systems, providing indoor wayfinding assistance in malls, airports, and other large venues [6].

AR has also been used in the entertainment and video game industry by blending virtual elements with the real world, AR has created a new dimension for interactive experiences. Mobile AR games like Pokémon Go have captured the attention of millions, by interacting with virtual characters. Live events and performances have also embraced AR, offering audiences unique and captivating experiences that blur the boundaries between physical and virtual elements.

The manufacturing industry has not been left behind with advances in augmented reality either. This has been achieved by optimizing production processes, training workers, and improving maintenance procedures; augmented reality has significantly increased efficiency and productivity. AR enables a work instruction guide, reducing errors and accelerating operations. Technicians, equipped with AR devices, can access real-time data and digital manuals overlaid onto machinery, facilitating repairs, and minimizing downtime. Digital twins, virtual replicas of physical assets, when combined with AR, enable operators to monitor and control equipment like in **Figure 1**, predict maintenance needs, and optimize overall performance. In general, AR enables having a better understanding of the process or machine or how it is really working.

The connection between cyber-physical systems, digital twins, and augmented reality has unlocked great potential in many industries. By the integration of AR with the Internet of Things (IoT) devices and sensors, real-time data can be overlaid onto physical objects, improving process monitoring, control, and decision-making.

According to the National Science Foundation, NSF, cyber-physical systems are devices capable of integrating computational algorithms and physical components, which allow to be equipped with storage and communication in order to control and interact with each other, surpassing the current integrated systems in terms of capacity, adaptability, scalability, resilience, security, and usability. This implies related technical challenges such as the development of control systems with self-learning capabilities and software development that serve as an interface between the physical and virtual system, also supported by the development of new technologies or new concepts such as the Internet of Things (IoT) [8].

Figure 1.
Digital overlay onto real machinery [7].

Cyber-physical systems and AR have the ability to relate to physical objects from a virtual environment in order to monitor and/or control. They use available information in the collection of data for the virtual world, being able to integrate in some cases automatic learning techniques and decision making.

The design and implementation of cyber-physical systems occupy an important part in the transition toward the fourth industrial revolution (Industry 4.0), which includes the digitalization, networks, and intelligence of the manufacturing industry [9]. **Figure 2** shows a general outline of the current framework of cyber-physical systems, focused on the integration, interconnection, and interaction of multiple layers or levels from a physical layer composed of all hardware elements such as sensors, machines, and robots to the most abstract layer to perform monitoring, control, and

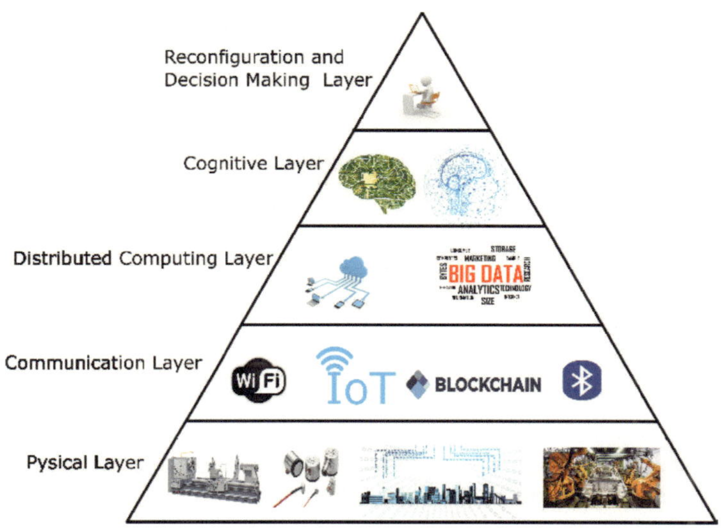

Figure 2.
Framework of cyber-physical systems [10].

self-reconfiguration functions, in order to cover the entire value chain in the industry of the future [10].

The advantages of cyber-physical systems and AR applications can be exploited in multiple applications such as manufacturing, energy, health, transport, smart cities, and so on. Below are examples of a new generation of development and solutions:

- Control of a machine tool or wind turbine to optimize its performance.
- Monitoring the status of the machine or system and optimizing its operation and maintenance strategy.
- Vehicles that communicate with others and with the road—infrastructure to determine the appropriate speed or routes while real time and augmented reality information is being displayed on the traffic windshield.

A digital twin is the digital replica or virtual representation of a machine in the physical world that simulates the behavior of its real counterpart. In this investigation, the virtual world representation of the real system seeks to control and monitor a centrifugal pump test bench through a web page.

Digital twins are a simulated process that uses real data from a physical model, such as sensors. In this way, they can reflect the life cycle of the process that corresponds to the physical equipment, thus analyzing various scenarios to improve the use of the machine or to be able to prevent failures.

An important requirement in the concept of digital twin is that it must be a dynamically and continuously updated representation of the actual product, device, or physical process. It should not be a static representation of real space. Real and virtual spaces are connected from manufacturing and operation to the disposal of the product, device, or process. Sensor information, user reports, and other information collected through manufacturing and operating processes must be continuously transferred to the digital twin. Predictions, control parameters, and other variables, which can be used to design and operate the real device, must be continuously transferred from the virtual space to the real space.

These are some of the biggest importance of digital twins in industry and pumping systems.

- Digital twins help test different approaches to reduce expenses without putting risk or cost at stake.
- They reduce engineering time, testing, commissioning, and upgrading costs of pumping stations to improve performance.
- The preliminary tests that can be carried out before the installation of the pumping stations can reduce the cost of commissioning and accelerate it, thus increasing reliability.
- Some of the problems such as leakage, water hammer, pump cavitation, or flow-induced vibration are problems that can be treated with a digital twin. At the right moment when the experimental data do not match the digital twin data, an alarm will be generated, alerting the operators that something may be wrong, thus preventing failures. This is how via constant monitoring the digital twin and the cyber-physical system can be linked together.

- One of the pillars of digital twins is to be able to monitor the condition of the process, generating alerts that can trigger preventive maintenance before problems occur; this is of paramount importance since the lives of the staff are at stake and could generate negative impacts on the environment. This information can also be relevant to the scheduling of the maintenance of a piece of equipment, since in this way, maintenance can be avoided ahead of time or worse, after it is needed. A just-in-time methodology could be managed, saving operating costs, spare parts, and staff time, increasing their productivity.

- During the manufacturing process of pumping systems, a digital twin provides relevant information about the performance of the equipment during the entire time of the process.

- In pumping systems, it is important to measure two types of pressure variables at the inlet and outlet of the turbomachine and the flow rate it delivers to the final discharge line. These are the instruments used for data collection and augmented reality visualization of the current value of sensors.

- Immersive applications can be created that represent scenarios that allow the worker to visualize a complete situation in the work environment, which teach about risk prevention and training to avoid making mistakes in practice.

- Through augmented reality, the real environment can be augmented with text, labels, documents, 3D models, and videos, which will lead to fewer errors and faster and higher quality of the service process.

A digital twin consists of 5 components:

1. The physical equipment.

2. The IoT that allows the communication of the information generated by the physical equipment.

3. Storage of information.

4. Analysis of such information. This stage is where the making of the best decisions that the process needs is promoted.

5. Equipment actuators. They allow the information generated and analyzed to meet the initial objective; they will no longer be only calculations and analysis but also actions that directly influence the team, optimizing the process and reducing costs.

2. Literature review

Virtual reality (VR) and augmented reality (AR) have both been employed in a range of educational environments [11], inclusive of: mathematics [12] and geometry [13], chemistry [14], biology [15], and mechanical engineering [16].

For instance, in [17], the authors consider experimental tests in order to validate augmented reality efficiency. They used a centrifugal pump as well. The authors paid attention to the fact that the oil companies in Russia suffered more than 4500 cases of downtime each year due to equipment failure. Repair costs exceeded 2.5 billion rubles. This also means that no raw material is extracted, which leads to a loss of 500,000 tons of oil, which is about 3.8 billion rubles. In Russia, the weather is something to consider; temperatures can reach as low as −50 degrees Celsius. They have to maximize automation and minimize workers. This can be solved with a digital twin that controls the operation on the cloud. The authors argue that VR simulators for training will enable specialists to have relevant equipment testing, perform manipulations without health risk, reduce training time by automating the process of tool operation, and apply various scenarios for the development of the trainee's skills and thinking strategies in out of the ordinary situations. They tested the disassembly of different parts of the pump, such as the coupling guard, the two halves of the clutch, and the pin, and dismantling of the engine section. The idea was to analyze the times spent by different groups. Some groups had only physical instructions of the process; others had help from an expert; others used the recommendations of the augmented reality system, and the last group used the software and the expert if it was necessary. The group that took the longest time to complete the tests was the group that conventionally worked in the industry, which is to use only the equipment documentation. Always scoring among the best were the teams that used the augmented reality technology when solving the tests. They proved the main hypothesis, which is that the AR system reduces the maintenance time of oil pumps.

In [18], this study proposes the creation of a virtual training system for the installation, calibration, and commissioning of HART transmitters in dynamic and potentially hazardous environments, such as oil and gas process plants. Virtual reality and augmented reality both are processes that optimize and reduce training time and costs. Through interactive AR simulations, trainees can familiarize themselves with the principles of fluid dynamics, pump operation, and pipeline management. By overlaying virtual components, such as pumps, valves, or pipelines, onto a physical training environment, trainees can gain practical experience in a controlled virtual setting. This immersive approach enhances learning outcomes and allows for hands-on practice before engaging with real-world equipment.

The promise of virtual environments in training has been demonstrated because they allow focusing on contemporary training needs and integrating multiple requirements for using field equipment in a single program. The use of VR systems in training procedures has proven to be a significant and useful technology. First and foremost, these technologies, when implemented, help to achieve the goal of reducing economic losses, since incorrect calibrations and configurations in real life would be avoided to a large percentage, besides avoiding the risk that humans present in the process for the oil and gas industries. Employee knowledge and performance is greatly improved through this type of training—an effective and affordable substitute for handling, training, and learning about industrial equipment. All of this is possible in these virtual training facilities. The use of this technology will not only help prepare employees to operate any equipment, but it will also ensure the safety of the user as well as the simulated equipment and process. When some technologies are used improperly and the inherent risks of the technology are present, it can result in accidents that can have a negative impact on the environment, as well as economic and personal losses.

In pipeline inspection and maintenance, by utilizing AR devices, field technicians can access real-time data overlaid onto the physical pipeline infrastructure. This enables them to swiftly identify potential issues, monitor pipeline conditions, and efficiently carry out maintenance tasks. With access to pertinent information such as pipeline parameters, maintenance history, and sensor data, technicians can make informed decisions, ensuring optimal performance and reducing downtime [19].

AR's ability to provide insightful visualizations of fluid flow within pipelines or industrial systems is another remarkable application. By overlaying virtual representations of fluid movement onto the physical environment, engineers and operators gain a deeper understanding of fluid dynamics. This facilitates the identification of potential bottlenecks, optimization of system design and operation, and effective troubleshooting of anomalies in fluid flow [20].

AR's remote assistance and collaboration capabilities have transformed the way field technicians and operators tackle complex tasks. Through AR devices, experts can provide real-time guidance and support from a remote location. By sharing their perspectives and overlaying annotations, instructions, or diagrams onto the technician's field of view, experts can assist in troubleshooting, repairs, and intricate operations related to fluid systems. This seamless remote collaboration reduces travel costs, minimizes downtime, and facilitates efficient problem-solving. Vuforia Chalk, **Figure 3**, is one of the most famous software related to this topic.

Furthermore, AR has found a vital role in safety training and hazard recognition within fluid-related industries. By simulating hazardous scenarios, such as leaks, pressure hazards, or chemical exposures, trainees can visualize potential dangers overlaid onto their physical surroundings. This immersive experience enhances hazard recognition, risk assessment, and emergency preparedness. Ultimately, this application of AR contributes to improved safety protocols, reducing the occurrence of accidents and promoting a culture of safety within the industry. "Safety

Figure 3.
Vuforia chalk software [21].

training and hazard identification: AR can enhance safety training programs by simulating hazardous scenarios and providing interactive training modules. It can also help identify potential hazards on-site by overlaying warning signs, safety guidelines, and visual cues onto the real environment, promoting a safer work environment" [22].

3. Materials and methods

The test bench of multiple centrifugal pumps in variable configuration of UNAB is composed of two Pedrollo brand centrifugal pumps, reference CP 620, pressure and flow sensors, a control panel where the data are acquired and the process variable is manipulated, a PLC S7–1200, a V20 frequency inverter, a CM 1241 module, and an SM 1231 module. An interactive model is required that allows the autonomous and efficient use of the test bench of multiple centrifugal pumps in variable configuration, providing clear instructions and the step by step for the handling of this. This will be worked through a local area network where you can also configure a VPN (if you have the permissions to make configurations on the edge router) to have remote access and be able to acquire data, perform maintenance, or even solve breakdowns, saving costs and time for the company at an industrial level. With augmented reality, the project will have highly visual content with which important digital information will be presented in the context of a physical environment; this allows students to connect and improve academic results; applying it will achieve an optimal way to easily create and distribute work instructions by overlaying digital content in the real world. The following subsections will cover an outline of the interconnection of the cyber-physical system and augmented reality on the test bench of two centrifugal pumps in variable configuration of UNAB.

The development of the digital twin for the pumping system under study was carried out according to the following methodology:

3.1 The 3D model

First, you must have or model the 3D model of the equipment in interest. In this project, the fluid test bench was modeled in the SolidWorks software. The 3D model was modeled at the scale of the real equipment, in such a way that the virtual and real model was approached on the largest possible scale to avoid having tracking and recognition problems when working with augmented reality (Vuforia Model Target) (**Figure 4**).

3.2 Augmented reality interface

The second step is to develop the augmented reality interface for the efficient and safe operation of the equipment. In this project, it was decided to work in unity with the Vuforia SDK where functionalities can be added such as monitoring the sensors, adding information about the pumping equipment, step-by-step instructions for the management of the system and the explosion of the components for an immersion in the operation of the machine (**Figure 5**).

Unity is a development engine or game engine. A game engine refers to software that has a series of programming routines that allow the design, creation, and

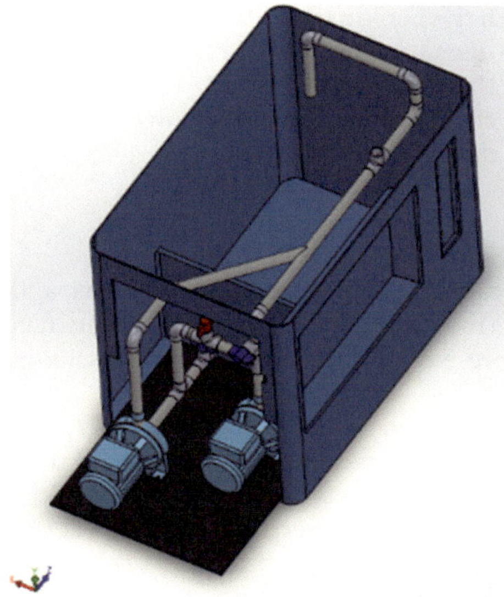

Figure 4.
The 3D model.

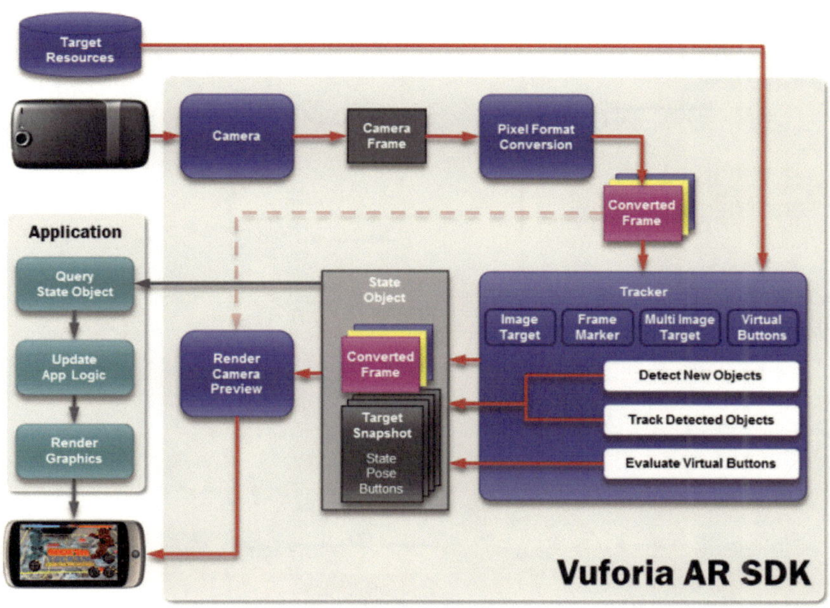

Figure 5.
Vuforia SDK [23].

operation of an interactive environment, that is, from a video game. Currently, more than 60% of all content developed with virtual reality and augmented reality is created with this multiplatform. The operation of Vuforia consists of sending captures of images from the camera to its servers and contrasting them with what exists in the database. At the moment when there is a match, it sends us an object with the metadata associated with the bookmark. If this procedure was done constantly, the resources that would be consumed would be enormous, so you have to say "When" you must scan to detect the marker.

These libraries were chosen mainly for their powerful algorithms that, being so developed, offer one of the best results today.

3.3 Target models

The most common tracking methods are images, areas, and targets tracking. The method used in this investigation was model tracking. It consists of recognizing objects by shape using pre-existing 3D models. **Figure 6** shows the steps for creating target models.

The model must be created in the MTG (Model Target Generator) software provided by Vuforia. A target model requires the user to hold their device at a particular distance and angle so that tracking a target model can be initialized. To help with this process, the application will draw the vertices of the object (guide views) so that by overlapping with the object in the real environment, the augmented reality experience can be initialized (**Figure 7**).

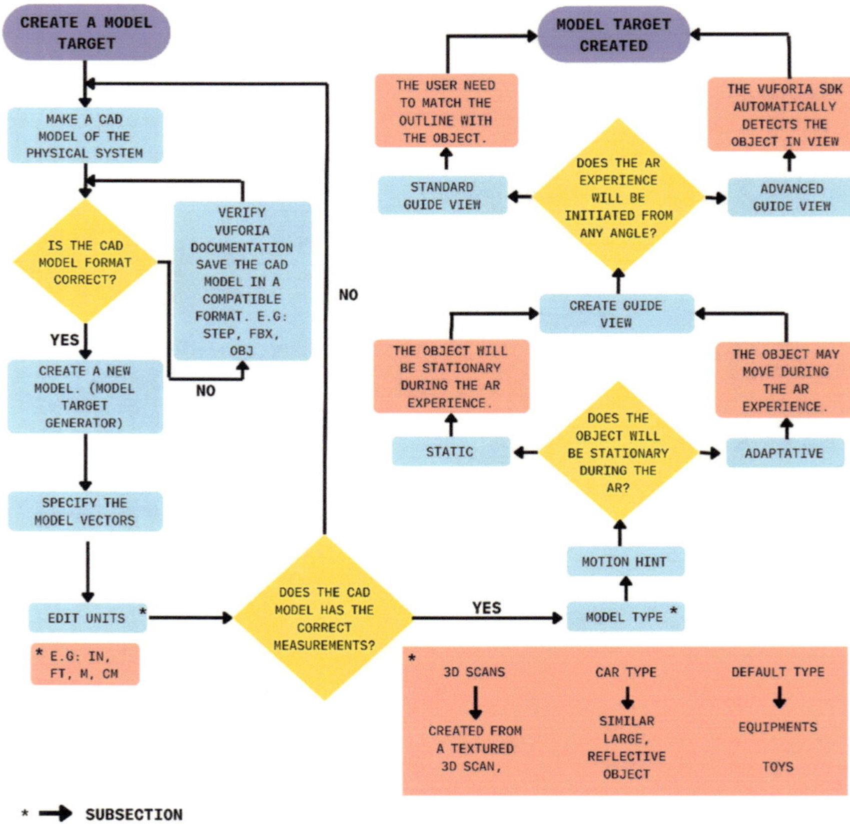

Figure 6.
Steps for target model creation.

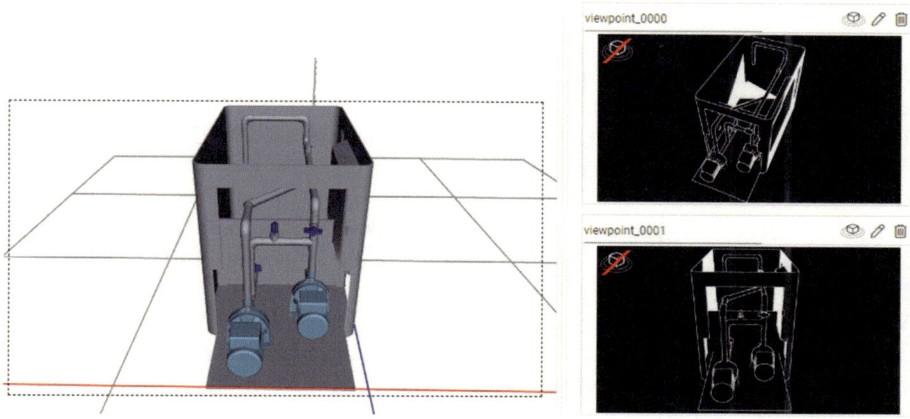

Figure 7.
Guide views in the MTG [24].

3.4 Automation and web development

Finally, the web application must be developed in which the functionalities of the control and monitoring of the fluid bank will be implemented. A local area network is also configured through which the current state of the sensors of the pumping equipment will be transmitted with the help of objects in unity such as "OnNewSearchResult"and "TargetSearchResult." "OnNewSearchResult" is an event that is handled when the Vuforia server returns a positive detection. The object returns a "TargetSearchResult," and the metadata variable has the metadata associated with the detected marker. After this, the monitoring of the object in the marker is enabled so that Vuforia's artificial vision algorithm tracks the marker (**Figure 8**).

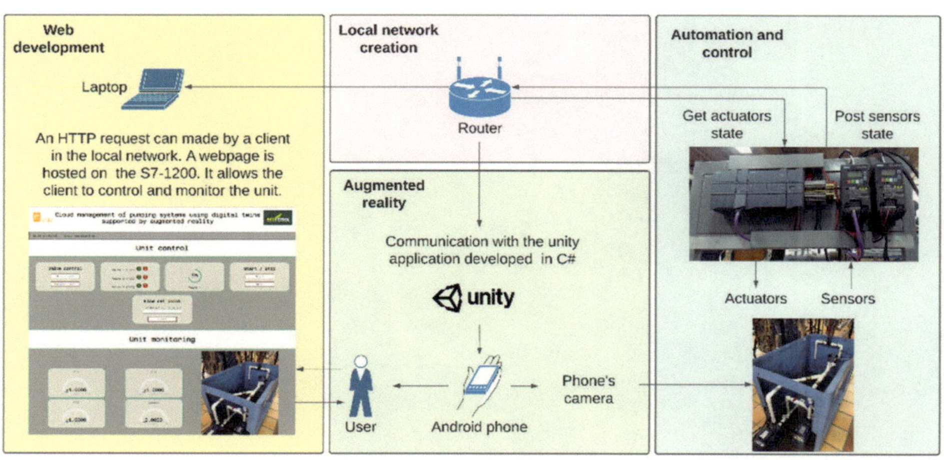

Figure 8.
Control and monitoring of the pumping system arquitecture.

3.4.1 AWP commands

Automation Web Programming (AWP) commands is a special command syntax for exchanging data between the CPU and the user page (HTML file).

AWP commands are entered in the form of comments in HTML and give you the following options for your user pages:

Read variables PLC. := < Varname>:

Write variables PLC. <!-- AWP_In_Variable Name = '<Varname1 > '-->

Read special variables. <!-- AWP_Out_Variable Name = '<Typ>: <Name>'-->

Write special variables. <!-- AWP_In_Variable Name = '<Typ>: <Name>'-->

The storage an analysis of the information is crucial in order to accomplish predictive maintenance or take the best possible choices for the process.

4. Results

Figure 9 shows the augmented reality interface with its buttons to enter each of the sections and monitor the process. In the pressure and flow section, you can access the instrument datasheet information along with related graphs. In the explosion section, you can see an example animation of how the parts can be exploded in augmented reality to get a better insight of the equipment. This is of vital importance for educational environments as well as to fully understand the operation of a process. The assembly section shows animations going from exploded view to normal view. The menu button takes you back to the main menu where you have more options in the application.

Figure 10 shows the frontend application. The process control system allows to control the flow rate and the start or shutdown of the machine. It also has a monitoring section, where the information from the four sensors is displayed in the form of a graph that allows an easy interpretation of what is happening with the test bench.

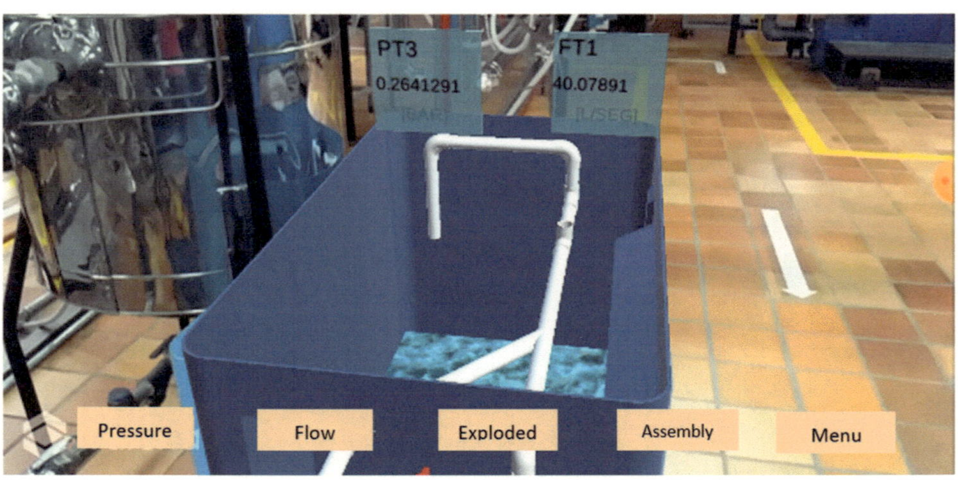

Figure 9.
Augmented reality interface.

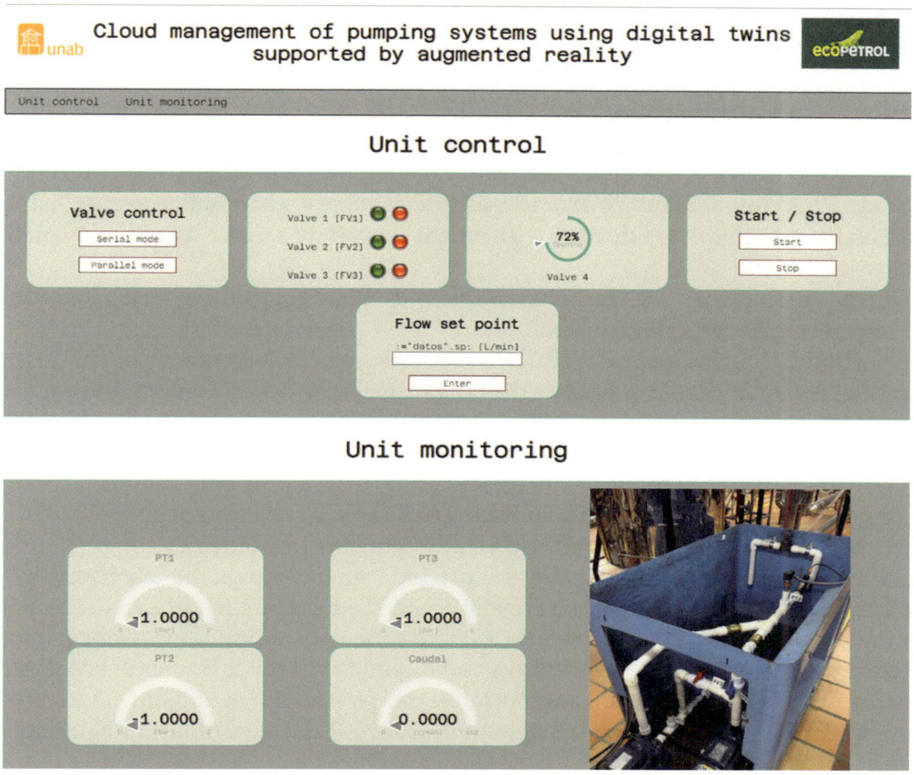

Figure 10.
Web interface.

5. Conclusions

In the field of centrifugal pumps, AR has emerged as a valuable tool for assembly, installation, and maintenance processes. Technicians equipped with AR headsets or mobile devices can leverage the technology to overlay step-by-step instructions and 3D models of pump components onto the physical pump. This visual guidance aids in accurate assembly, mitigates errors, and simplifies maintenance procedures, enhancing overall operational efficiency.

The integration of augmented reality in fluid-related industries exemplifies the transformative potential of this technology. By providing real-time information, visualization capabilities, training simulations, and remote collaboration tools, AR optimizes fluid transport, pipeline management, and centrifugal pump operations. With its ability to enhance efficiency, improve safety, and facilitate informed decision-making, AR continues to reshape these industries, paving the way for a more efficient and sustainable future.

This development improves efficiency, reduces workers' risk, and improves their productivity.

Augmented reality can enhance the capabilities of apprentices, offering additional information when doing their internships. In addition, they have a wow factor that captures the attention of employees much more effectively.

It provides clear, step-by-step instructions for handling any pumping system or process; accelerates training; transfers expert knowledge to new generations of engineers; and accelerates the learning curve for newly hired engineers through interactive information in augmented reality.

In pumping system equipment, there is a vast amount of technical information, and being able to see it in augmented reality reduces the time required to interpret manuals. This complete information is easily accessible with a simple gesture, which is particularly valuable when dealing with extensive manuals that may not be available on site.

This technology has been shown to improve attention, promote lasting knowledge. and more effectively explain difficult topics.

The digital twin is viable as a design methodology because it can be based on the current product to be able to predict the operation of the device under different scenarios and configurations, optimize its conditions, and evaluate changes effectively without resorting to iterative prototyping investments.

Over time, the value of highly integrated, data-driven pumping solutions will become increasingly apparent and indispensable.

In conclusion, augmented reality is a transformative technology with boundless applications across various industries. Its ability to seamlessly blend virtual and real-world elements has revolutionized education, healthcare, manufacturing, retail, entertainment, and more. As AR continues to advance, its integration with cyber-physical systems and digital twins holds immense potential for reshaping how we interact with and optimize our physical environment. The future of augmented reality is bright, and the possibilities are at the edge of our imagination.

Acknowledgements

The authors acknowledge the support from Universidad Autónoma de Bucaramanga and ECOPETROL to carry out this work.

Conflict of interest

The authors declare no conflict of interest.

Author details

Juan David Rueda Martínez[1], Gonzalo Andrés Moreno Olano[1], Sebastian Roa-Prada[1*], Hernán Darío Mantilla Hernández[2] and Nestor Fernando Saavedra Trujillo[2]

1 Universidad Autónoma de Bucaramanga, Colombia

2 ECOPETROL-ICP, Piedecuesta, Colombia

*Address all correspondence to: sroa@unab.edu.co

References

[1] Laverde JSR. Herramienta pedagogica utilizando realidad aumentada para el apoyo en la enseñanza de ciencias naturales enfocada a estudiantes de grado sexto [thesis]. Bucaramanga: Universidad Autonoma de Bucaramanga; 2019

[2] Barcali E, Iadanza E, Manetti L, Francia P, Nardi C, Bocchi L. Augmented reality in surgery: A scoping review. Applied Sciences. 2022;**12**:6890. DOI: 10.3390/app12146890

[3] Augmented Reality Can Be Real Gucci [Internet]. 2021. Available from: https://shorturl.at/jlwLY [Accessed: June 12, 2023]

[4] Say hej to IKEA Place [Internet]. Available from: https://shorturl.at/azEK1 [Accessed: June 12, 2023]

[5] How Augmented Reality Can Revolutionize Political Campaigning [Internet]. 2020. Available from: https://shorturl.at/deKOV [Accessed: June 14, 2023]

[6] Bringing your map to life, one image at a time [Internet]. Available from: https://www.google.com/intl/en-419_co/streetview. [Accessed: June 14, 2023]

[7] Augmented Reality Applications [Internet]. Available from: https://www.howden.com/en-us/whitepapers/augmented-reality-applications [Accessed: June 15, 2023]

[8] Cyber-Physical Systems (CPS) [Internet]. Available from: https://shorturl.at/bBLMR [Accessed: May 11, 2023]

[9] Javaid M, Haleem A, Singh RP, Suman R. An integrated outlook of cyber–physical Systems for Industry 4.0: Topical practices, architecture, and applications. Green Technologies and Sustainability. 2023;**1**(1):100001. DOI: 10.1016/j.grets.2022.100001

[10] Villalonga F, Castaño G, Beruvides R, Haber SS, Kossakowska J. Visual analytics framework for condition monitoring in cyber-physical systems. In: 2019 23rd International Conference on System Theory, Control and Computing (ICSTCC). Sinaia, Romania: IEEE; 2019

[11] Towey D, Walker J, Austin C. Developing virtual reality open educational resources in a Sino-foreign higher education institution: Challenges and strategies. In: IEEE International Conference on Teaching, Assessment, and Learning for Engineering (TALE), Wollongong, NSW, Australia. 2018. pp. 416-422

[12] Keblitchi M, Hinmi A, Bai H. The effect of modem mathematics computer games on mathematics achievement and class motivation. Computers & Education. 2010;**55**(2):427-443

[13] Hwang W-Y, Hu S-S. Analysis of peer learning behaviours using multiple representations in virtual reality and their impacts on geomeüy problem solving. Computers & Education. 2013;**62**:308-319

[14] Merchant Z, Goetz E, Keeney-Kennicutt W, Kwok O, Cifuentes L, Davis TJ. The leamer characteristics, features of desktop 3D viltual reality environments, and college chemist1Y instmction: A stmctural equation modelling analysis. Computers & Education. 2012;**59**(2):551-568

[15] Lee EA, Wong KW, Fung CC. How does desktop virtual reality enhance leaming outcomes? A stmctural

modelling equation approach. Computers & Education. 2010;**55**:1424-1442

[16] Coller BD, Shemoff DJ. Video game-based education in mechanical engineefing: A look at student engagement. International Journal of Engineering Education. 2009;**25**(2):308-317

[17] Koteleva N, Buslaev G, Valnev V, Kunshin A. Augmented reality system and maintenance of oil pumps. The International Journal of Engineering. 2020;**33**(8):1620-1628. DOI: 10.5829/ije.2020.33.08b.20

[18] Garcia CA, Naranjo JE, Ortiz A, Garcia MV. An approach of virtual reality environment for technicians training in upstream sector. In: The International Federation of Automatic Control. Bilbao, Spain: Elsevier; 2019. DOI: 10.1016/j.ifacol.2019.08.222

[19] Shekargoftar A, Taghaddos H, Azodi A, Tak AN, Ghorab K. An integrated framework for operation and maintenance of gas utility pipeline using BIM, GIS, and AR. Journal of Performance of Constructed Facilities. 2022;**36**(3). Abstract. DOI: 10.1061/(ASCE)CF.1943-5509.0001722

[20] Mourtzis D, Angelopoulos J, Panopoulos N. Challenges and opportunities for integrating augmented reality and computational fluid dynamics modeling under the framework of industry 4.0. Procedia CIRP. 2022;**106**:215-220. DOI: 10.1016/j.procir.2022.02.181

[21] PTC VUFORIA. Vuforia Chalk: Remote Assistance Powered by Augmented Reality [Internet]. Available from: https://www.ptc.com/es/products/vuforia/vuforia-chalk. [Accessed: June 20, 2023]

[22] Abd El-Rahman Samy Bo Shaieb. Augmented reality applications in the fields of civil engineering. The International Journal of Advances Engineering and Civil Research. 2022;**2**(2):64-93

[23] Shaaban O, Mat RC, Hafiz M. The development of mobile augmented reality for laptop maintenance (MAR4LM). Jurnal Teknologi. 2015;**77**:91-96. DOI: 10.11113/jt.v77.6842

[24] PTC VUFORIA. PTC: Augmented Reality [Internet]. Available from: https://www.ptc.com/products/vuforia. [Accessed: June 21, 2023]

Section 3

Augmented Reality in Medicine

Chapter 9

Current Status and Future Perspectives for Augmented Reality Navigation in Neurosurgery and Orthopedic Surgery

Quentin Neuville, Thierry Scheerlinck and Johnny Duerinck

Abstract

Augmented reality (AR) for surgical navigation is a relatively new but rapidly evolving and promising field. AR can add navigation to the surgical workflow in a more intuitive way, improving ergonomics and precision. Neurosurgery has a long tradition in computer-assisted surgical navigation and was the first discipline to use this technology to navigate interventions. Orthopedic surgery is following this path with a wide range of new use cases currently in development. In this chapter, we will describe the evolution of AR as a surgical navigation tool, focusing on application developed for neurosurgery and orthopedic surgery. Based on our own experience, we will also discuss future perspectives and the hurdles to be overcome before the widespread breakthrough of this technology.

Keywords: augmented reality, AR, neurosurgery, orthopedic surgery, navigation

1. Introduction

Due to limited imaging possibilities in the early days of surgery, large incisions were often made to expose as much normal anatomy as possible for orientation purposes. This led to long and more extensive interventions with steep learning curves. It also resulted in more patient discomfort, a higher likelihood of complications and longer hospital stays. The advent of intra-operative imaging modalities such as fluoroscopy, CT and MRI, but also computer-assisted navigation (CAN) systems, have been milestones towards more accurate and minimally invasive surgery. CAN systems were originally developed for brain surgery, this is why they are often referred to as neuronavigation systems. Nowadays, CAN systems present a wide range of applications in different surgical disciplines, such as neurosurgery, orthopedic surgery, oral and maxillofacial surgery and otolaryngology. However, despite the widespread availability of CAN systems in North America and Europe, on average only 11% of surgeons use it routinely [1]. This is attributed to some fundamental shortcomings including, first of all, the bulkiness of the CAN devices. Current CAN devices take up a lot of space in the operating theater, as they require a computer system, external tracking camera and one or more screens [2]. Secondly, the systems require a time-consuming set up including a patient registration procedure [3]. A third drawback is that the surgeon is typically looking at the screen, which shifts his attention away from the patient and can present a potential hazard when instruments - such as the navigation pointer itself - are held in the operative field [4, 5]. Fourth, the outside-in tracking by an external optical tracking camera (mostly infrared camera) often creates line of sight (LOS) interruption when people or objects in the operating room enter in the line of sight of the camera [6, 7]. The significant cost of these systems might also present a hurdle holding surgeons back from fully adopting computer-assisted navigation for their surgical interventions [8, 9].

Augmented reality is a technology that superimposes computer-generated information or images on a user's view of the real world. Unlike virtual reality (VR), AR does not occlude the real environment, but rather overlays virtual information on it. Therefore, AR based navigation is a promising technique for open and minimally invasive surgical procedures, as it does not block the surgical field but can project additional data and/or hidden anatomy on top of it. In order to turn an augmented reality device into a navigation system, three additional steps are needed: calibration of the AR device, tracking and image-to-patient registration, which matches the AR models created from pre-operative segmentations of medical data to the intra-operative object [10, 11]. Calibration of the AR device is necessary to determine the AR display coordinate system in relation to the outside world. During tracking, we determine the real-time position and orientation of the patient and surgical instruments in relation to the AR device. It is necessary to maintain good alignment between the virtual and physical environment and allow accurate navigation. Augmented reality (AR)-based navigation has the potential to solve some of the shortcomings of conventional CAN and thus to increase the use of navigation in routine interventions. In addition, AR-based navigation can be complementary to commercially available CAN systems in procedures where CAN is routinely used. The goal of this chapter is to discuss the Augmented Reality (AR) technologies that are developed for neurosurgery and orthopedic surgery, to understand the current evidence regarding their benefit, to consider challenges limiting implementation in clinical practice and to get an idea about the future perspectives of this novel technology.

2. What types of AR-based navigation are currently being developed for neuro- and orthopedic surgery

There is significant variability in the way AR-based navigation is being deployed, although all attempt to merge imaging data to the surgical field. However, not all systems try to achieve this based on calibration, tracking and registration described previously.

2.1 Types of visualizations

There are 3 main techniques of displaying the AR models for surgical guidance: (i) on a head mounted display (HMD) [8, 12, 13] (ii) on a projector [14], and (iii) on an external display [15]. In our opinion, both techniques based on a projector and on external displays such as phone, tablet, screen are suboptimal. They maintain many of the drawbacks of commercially available navigation systems discussed earlier. Therefore, we prefer using an HMD, which is the only method capable of visualizing the AR in 3D on top of the patient's anatomy. As such, surgeons can view holograms and the surgical field simultaneously, without lag or attention shifts.

2.2 Types of registration

One of the key issues of AR-based surgical navigation is the image-to-patient registration accuracy. Although the requirements depend on the type of surgical intervention, it is essential to maintain an accurate image registration throughout the whole procedure. The most commonly used registration methods for AR in neurosurgery and orthopedics are manual registration, point-based registration and surface registration. During manual registration, the surgeon positions the virtual object visually on top of the physical object. This is the least accurate registration method and it is not suitable for surgeries requiring high accuracy or for minimally invasive procedures. In point-based registration, a specific algorithm looks for the best fit between a set of preoperatively defined points in the image dataset and corresponding anatomic intra-operative points [16]. Surface-based registration is a more sophisticated and accurate version of point-based registration and is performed in two steps [15]. First, a rough initial alignment is obtained using point-based registration methods. In a second step, a more dens point-cloud or surface mesh is obtained and matched to the geometric shapes of the preoperative model using a surface-based alignment process, such as an iterative closest point (ICP) algorithm. Most commercially available navigation systems for neurosurgery and orthopedic surgery use this technique as it provides a practical and accurate alignment. In order to collect intra-operative points, the surgeon needs a tracked pointer i.e., a stylus that can be tracked by the AR device and that allows to define 3D coordinates of points in the real world. In the future, automatic point-cloud or surface mesh collection could be combined with the extraction of anatomical landmarks based on deep learning methods. When performed in real time, this could replace manual point-based registration methods and accelerate the registration process without need for manual interaction [15]. Manual registration can reach accuracies of 4–6 mm while point-based registration and surface-based registration reach accuracies of 2–3 mm [16–21]. For point- and surface-based registration, the accuracy depends on the number of points collected and the distribution of the collected points relative to the target [22].

2.3 Types of tracking

After registration, the virtual objects and patient anatomy are aligned. However, to maintain the alignment, movements of the patient and/or the AR device must be calculated and compensated for in real-time. During AR-guided surgery, this can be based on the Simultaneous Localization and Mapping (SLAM) tracking system that is integrated in the AR-HMD or on images from other onboard and/or external cameras. Most commercially available AR-HMD's, including the HoloLens II (Microsoft, Redmond, USA) come with SLAM tracking. This is a self-localization technology, that captures changes in the surrounding with the integrated camera and displays virtual 3D image in a fixed relation to the real world [15]. However, SLAM tracking can result in an important drift of the virtual objects and the tracking accuracy of is too low for surgical use. For example, when the observer moves, the HoloLens has a mean perceived drift between 4.39 and 6 mm [19, 23]. Therefore, several research groups refine SLAM tracking with information of internal or external cameras to track markers in the surgical field. In most cases, these markers are tracked with the RGB camera (e.g., Vuforia). This results in a mean drift of 1.41 mm, which is a 68% improvement compared to SLAM tracking [19]. However, after experimenting with that method, we felt it was impractical and lacked accuracy for medical navigation applications. Tracking based on infrared cameras and retroreflective spheres - which is the method used by most commercially available CAN systems - is more accurate (mean shift 0.809 mm to 2.3 mm and angular errors of 1.038°) [13, 24, 25]. Most research groups use an external IR camera that presents the same shortcomings as commercially available systems. This includes potential LOS interruptions, the need for large and expensive external devices and lack of usability outside the OR. In addition, when using an external camera, the AR device itself must be equipped with markers so that its position can be monitored.

2.4 How we do it

We developed an IR tracking method based on the built-in IR camera of the HoloLens II. This "inside-out" tracking method allows tracking of IR reflective markers, bypassing the need for an external camera (**Figure 1**). The tracking and registration accuracy of our setup was first evaluated using a phantom head. In total, 20 phantom registrations were performed in an operating room setting. Ten registrations were performed with the AR-HMD, while 10 were carried out using a conventional neuronavigation system (Brainlab Curve 2; Brainlab AG, Munich, Germany). Registration errors remained below 2.0 mm and 2.0° for both AR-HMD based navigation and the conventional neuronavigation, with no significant difference between both systems [18]. Tracking from inside the AR device resolves some important drawbacks of commercially available navigation systems. First, tracking from inside the AR device avoids the need for the AR device to be equipped with a marker frame. Second, this approach prevents external LOS interruption. Third, the AR solution provides a fully integrated, mobile and ergonomic navigation setup that can easily be used in- or outside the operating theater. As such, it has the potential to

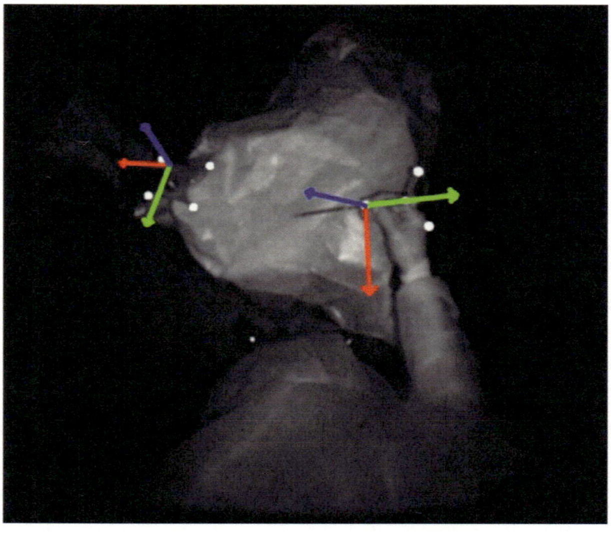

Figure 1.
Tracking of IR reflective markers from the HoloLens II.

increase the use of navigation and thus the accuracy of many routine procedures that are currently performed without navigation.

3. Applications in neurosurgery and orthopedic surgery

Most data regarding AR-based navigation are based on pre-clinical experiments on phantoms and cadavers or on small proof-of-concept trials. In the field of neurosurgery, experiments for AR-based navigation focused on identifying lesions, guiding tumor resection, guiding cranial biopsy, guiding external ventricular drain insertion and for preoperative planning of skin incisions and craniotomies [2, 15, 18, 26, 27]. In spine surgery, AR is mainly focusing on guiding pedicle screw placement and vertebroplasty [5, 28–32]. In orthopedics, AR has been described for trauma reconstruction, osteotomy, intramedullary nailing, arthroscopic surgery, oncology, K-wire implantation and arthroplasty [12, 20, 24, 33–37].

High-quality clinical data on the use of AR-navigation during surgery is currently lacking. Only a few clinical trials have been conducted to examine the accuracy of these systems in clinical practice and/or compare the use of an AR-based navigation system with conventional commercially available navigation systems for neurosurgical or orthopedic procedures [5, 8, 20, 29, 38–42]. In our view, the main advantage of AR navigation is that it allows introducing navigation in procedures (or specific phases of an intervention) where navigation is currently not used or where it is performing sub-optimally, as for example, with external ventricular drain placement or craniotomy planning.

3.1 Benefits of AR-based navigation

3.1.1 Improved accuracy of the surgical intervention

The introduction of dedicated, ergonomic, easy to use and portable AR-based navigation solutions could promote its use in- or outside the operating theater. As such, it could increase the accuracy of a large number of routine procedures that we perform today without the help of navigation. Nevertheless, many research groups focus on the use of AR-based navigation for interventions that are currently performed with CAN systems. This allows comparing their respective accuracies. Our consortium compared an internally developed AR-HMD based system for intracranial tumor resection planning to the Brainlab Curve 2 system (Brainlab AG, Munich, Germany). Image-to-patient Registration errors remained below 2.0 mm and 2.0° for both AR-HMD based navigation and the conventional neuronavigation, with no significant difference between both systems. However, for delineation of the tumor margins on the patients' skin in order to plan the incision and craniotomy, the AR-HMD based system was found to be superior in 65% of cases [18].

In a study evaluating AR for spinal pedicle screw placement, Molina et al. placed a total of 113 implants percutaneously (93 pedicle screws and 20 Jamshidi needles) in five cadavers using the XVision AR-HMD (Augmedics Ltd., Chicago, IL). The study reports an overall accuracy of 99.1% (Gertzbein-Robbins Grade A or B; <2 mm pedicle breach) with only one medial pedicle breach in a thoracic vertebra (Gertzbein-Robbins grade C; >2 mm pedicle breach) [9]. They compared their results to the literature of manual and robotic CAN as well as freehanded techniques. Overall, AR navigation was non-inferior to CAN and superior to free-hand procedures. In a proof-of-concept trial, Molina et al. reported the first in-human use of the same system to insert six pedicle screws. A 78-year-old female underwent an L4-S1 posterior lumbar interbody fusion. Clinical accuracy was 100% with a mean linear deviation of 2.07 mm and an angular deviation of 2.41° [8]. In a retrospective study comparing AR-guided and freehanded pedicle screw placement, Elmi-Terander et al. found that the percentages of clinically accurate screws (Gertzbein-Robbins grade A and B) were significantly higher in the AR-guided group compared to the freehand group (accuracies 93.9% and 89.6% respectively ($p < 0.05$)) [39]. Other trials confirmed that AR-guided pedicle screw placement has a similar accuracy compared to CAN guided pedicle screw placements and an improved accuracy compared to the free-hand technique [5, 20, 29, 41].

Researchers focusing on orthopedic procedures report similar findings during total knee arthroplasty (TKA), total hip arthroplasty (THA) and periacetabular osteotomy. During TKA, Tsukada et al. compared the femoral cut accuracy in 31 patients undergoing an AR-guided procedure, to a cohort of conventional TKA with an intramedullary guide. Using AR navigation, the coronal alignment was more accurate than the intramedullary guide [43]. Another prospective randomized controlled trial compared an AR-based portable hip navigation system to the portable HipAlign (OrthoAlign Inc., Aliso, USA) system to control cup version during THA. These authors report no differences between both systems in terms of cup inclination and only minor differences in cup anteversion [44]. Finally, Kiarostami et al. found that AR guided periacetabular osteotomies were more accurate compared to freehand procedures. The effect was more prominent for less-experienced surgeons [45].

In general, we conclude that, compared to freehanded procedures, AR based navigation systems are more accurate during both, neurosurgical and orthopedic procedures. However, compared to CAN systems, the accuracy is similar as demonstrated during pedicle screw placement as well as total hip and total knee arthroplasties.

3.1.2 Decrease of surgical time and radiation exposure

Augmented reality has the potential to shorten surgical time and reduce radiation exposure during procedures that are traditionally guided by fluoroscopy [32, 46, 47]. Trauma and minimally invasive surgery often require repeated fluoroscopic imaging to guide the intervention. Acquiring adequate and reproducible views comes at the cost of increased surgical time and radiation exposure. For this, Unberath et al. proposed an AR-based solution for C-arm repositioning. In a phantom experiment, mimicking pelvic trauma surgery, repositioning of the C-arm based on AR-guidance led to a significant reduction in surgical time and radiation dose compared to a trial-and-error approach [47]. In another setting, AR navigation of percutaneous vertebroplasty was compared to a fluoroscopy-guided procedure in nine patients. Here, both surgeon radiation exposure and surgical time were reduced, but accuracy also improved [48]. Similar results were found in phantom trials on AR-guided insertion of distal locking screws following intramedullary nailing and AR-guided K-wire insertion [34, 36].

3.1.3 Benefits for teaching

Several researchers also emphasized the importance of AR in teaching and reducing the learning curve for surgical skills acquisition [41, 49]. Our consortium compared the effect of AR-guidance and standardized training on the quality of external ventricular drain placement on a phantom model, by medical students. Our results showed a significantly higher number of good EVD placements (modified Kakarla scale grade 1) resulting from AR guidance, but not from training, in direct comparison to the untrained freehand performance. Training improved the accuracy of EVD placement in the freehand group, but not in the AR-guided group. Another study found that based on AR, medical students were able to place acetabular cups without supervision as accurately as they did when receiving hands-on instruction from an expert [50]. In addition, the technology can be used to visualize the surgical anatomy in 3D and to practice the course of the procedure in advance.

3.2 Limitations

Several barriers have hindered the adoption of AR-HMD based navigation in daily surgical practice. First, because of a lack of high-quality clinical studies, there are concerns about the accuracy of AR-based navigation during a surgical procedure Moreover, AR navigation validated in vitro, could be difficult to implement in clinical practice as strong scialytic light could interfere with AR tracking and visualization. Second, AR-based navigation is still a new technology in surgery and therefore it is currently not well integrated into the clinical workflow. To convince surgeons to use AR-HMDs in their daily practice, specific workflows must be developed integrating this technology within well-established clinical pathways. Third, for each surgical procedure, careful consideration must be given to which steps of the procedure the AR-HMD could be an added value. This will influence how and when it will be implemented, where and when the trackers will be placed, how the image-to-patient registration will be done and what should and should not be visualized. At some stages of the procedure, the surgeon may want to see both 3D segmentations of the patient anatomy and his pre-operative planning. While during other stages, it may be useful to visualize only a planned trajectory or a target to avoid interference of auxiliary information. Therefore, the surgeon needs the possibility to switch between various visualizations. As sterility is an issue, simple voice commands or hand gestures seems most appropriate for this, but it needs specific programming and training. Fourth, concerns do exist about the clarity and contrast of AR images and whether they are disruptive in the surgical setting [51]. The brightness of the AR overlay should be adjustable depending on the ambient light to prevent unintentional blindness. Moreover, overlaid images could distract the surgeon from important surgical events such as bleeding or unexpected objects entering the surgical field. Another important limitation is the fact that most commercially available HMDs – like the HoloLens – are designed to have focal lengths ranging from 2 meters to infinity, while surgeons typically work at a much closer distance. Therefore, the surgeon's eyes have to accommodate constantly when trying to view the view the physical and virtual world simultaneously [52, 53]. This could be challenging for surgeons wearing glasses and could be overcome by developing dedicated surgical headsets. Currently, several medical companies are working on specific AR-HMDs designed to work at close distances, such as the Magic Leap from Brainlab or the CART 3D from AR Spectra. Finally, multiple studies highlighted the learning curve associated with the introduction of AR-based navigation [50, 54]. However, this should not prevent the surgeon from evaluating and adopting this technology.

4. Workflow: how we do it

4.1 External ventricular drain placement

External ventricular drain (EVD) placement is a routine but life-saving procedure in neurosurgical practice. Placement is most often performed in the emergency department or the intensive care unit using a freehand technique. Despite it being a frequent procedure, the accuracy rate is only around 80%, with complications occurring in up to 40% of the cases [55]. Therefore, our consortium developed a HoloLens II application for AR-guided EVD placement, based on our internally developed inside-out IR tracking software [26]. Respecting the established workflow of EVD placement, we designed semi-automated image processing and image-to-patient registration algorithms, so that the AR application provided a smooth, completely mobile and highly accurate navigation setup to assist in EVD placement. This setup was tested in a phantom experiment to examine the impact of AR guidance on the accuracy and learning curve of EVD placement compared with the freehand technique (**Figure 2**). Sixteen medical students were randomly allocated to either the freehand technique group or the AR-guided group. Both groups were asked to place 4 EVD's (left and right on 2 phantom heads). The freehand group had access to pre-operative imaging but afterwards had to place the drain freehanded without guidance. The AR-guided group used the application on the AR-HMD, which provided an overlay of the virtual anatomical model and surgical plan on top of the phantom head. Next, both groups received training on how to place EVD's and were asked to repeat the same process. In total, 128 EVD's were placed. Both AR-guidance and training significantly improved the accuracy of the drain placements compared to the untrained freehand placement (**Figure 3**). The quality of EVD placement as assessed by the modified Kakarla scale (mKS) was significantly impacted by

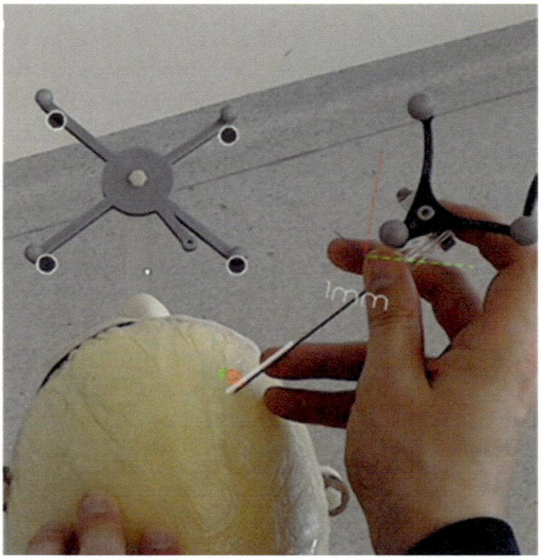

Figure 2.
Here you see the AR overlay provided for EVD placement. There is a Bullseye for the entry at Kocher's point in red (target) and the tracked EVD within a 2-mm threshold in green. Figure adopted from Van Gestel et al. [26].

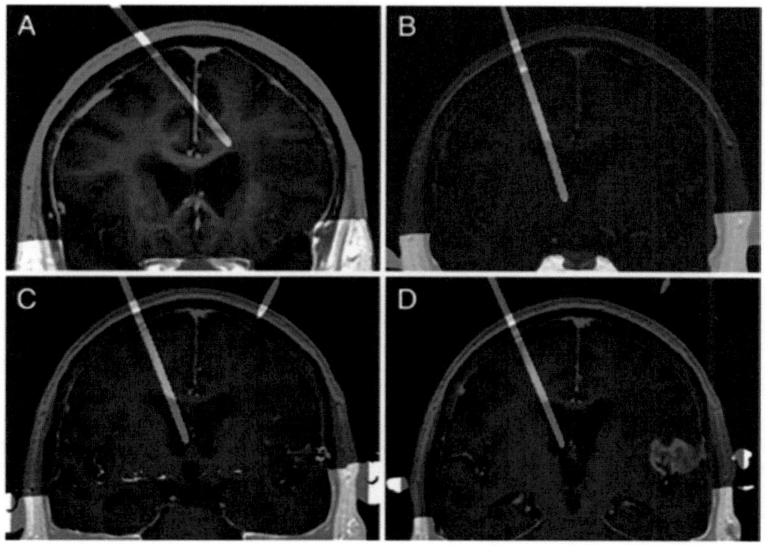

Figure 3.
Here you see an example of EVD placement results in the phantom. To appreciate the results, the post-EVD placement CT scan of the phantom is fused with the original brain MRI on which the trajectories were planned. Figures A and B represent freehand performances of a student before and after receiving standardized training respectively. Figures C and D represent AR-guided EVD placements before and after the standardized training respectively. Figure adopted from Van Gestel et al. [26].

AR-guidance ($p = 0.005$) but not by training ($p = 0.07$). Both AR-guided placements (before and after training) (59.4% mKS grade 1 for both) were significantly better than the untrained freehand performance (25.0% mKS grade 1). With AR-guidance, untrained students performed as well as trained students, which indicates that AR guidance not only improved performance but also positively impacted the learning curve [26].

4.2 Craniotomy planning

Careful planning of the skin incision, craniotomy and surgical approach are critical to successfully resect an intracranial lesion. Nowadays, CAN systems are indispensable for intracranial tumor resections. After image-to-patient registration, the surgeon uses a tracked pointer to indicate his entry point. Based on the tracked pointer, he then draws the tumor outline on the patient's skin to serve as a guide for skin incision and craniotomy. However, especially for deep-seated tumors, interpretation errors can lead to important deviations. Our consortium developed a HoloLens II based application to help plan tumor resection. After image-to-patient registration based on a pointer tracked with the inside-out IR-tracking software, the AR-HMD allows displaying the tumor outlines as well as critical structures on top of the patient's skin. We evaluated the accuracy and efficiency of this system in a prospective clinical trial. Surgeons and trainees with varying degrees of experience delineated the tumor outlines on the patient's skin, consecutively using conventional neuronavigation and the AR-HMD (**Figure 4**). In total 20 patients were included in the study. We found that AR-guided tumor delineation was deemed superior in 65% of cases, equally good in 30% of cases, and inferior in 5% of cases when compared to our conventional navigation system (Brainlab Curve 2; Brainlab AG, Munich, Germany).

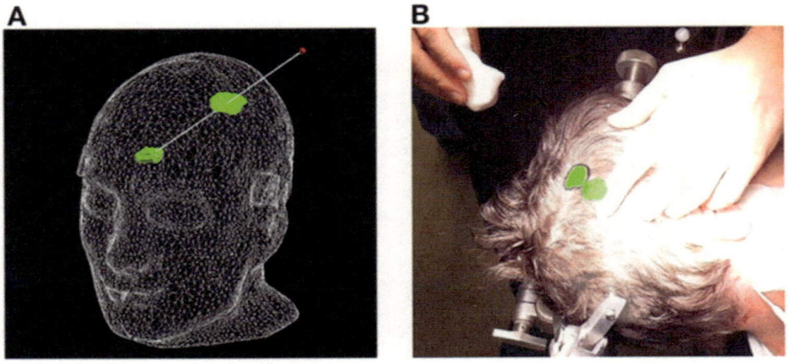

Figure 4.
Overview of the proposed AR-based system. (A) Phantom model with a deep-seated lesion (green) with the outlines projected on the skin (green outline) nearest to the tip of the handheld stylus (red dot). The white line illustrates the trajectory between the stylus' tip and the lesion. It is shown in this picture solely for illustrative purposes. (B) Is a view from the HoloLens II in a clinical case. The patients' tumor is displayed in AR on the correct position inside his head, together with the outlines orthographically projected on his skin, allowing the surgeon to delineate the tumor margins and plan skin incision (black marker). Figure adopted from Van Gestel et al. [18].

Moreover, the image-to-patient registration and planning of the surgical approach based on the AR-HMD significantly reduced the required time by 39% [18].

4.3 Hip center of rotation

In total hip arthroplasty (THA), restoring hip biomechanics is key to warrant a good functional result. Restoring the original femoral and acetabular center of rotation (COR) is a first important step to restore the original leg length, muscle tension and abductor level arm [56, 57]. Overall, there is agreement that following THA, the hip rotation center should be positioned within 5 mm of its anatomic location [58, 59]. Achieving that goal in a systematic way remains challenging [60, 61]. As mentioned earlier, the use of navigation has not yet found its way into daily practice of THA surgery, so in most cases, intra-operatively finding and restoring the original hip rotation center relies on surgeon experience. Our consortium developed an application on the HoloLens II to determine and render in AR the functional center of rotation (FCOR) of a phantom hip joint consisting of 20 cadaveric femurs and 3D printed acetabular cup analogues (**Figure 5**). Both the femurs and acetabular cups were equipped with a tracker. This hip phantom was CT scanned and a segmentation was made in 3D slicer to obtain the ground truth centers of rotation of the femoral heads and the cups. Next, two observers rotated the 20 cadaver femurs twice in its matching 3D printed cup, producing 80 measurements. Based on the displacement of the femoral tracker to the acetabular tracker, the inside-out IR tracking algorithm collected a point cloud (**Figure 6**). Through a pivot-fitting algorithm, the FCOR was then determined based on this point cloud and visualized in AR on top of the hip phantom. In our phantom trial, determination of FCOR through the proposed AR method resulted in an absolute error of 2.9 ± 1.4 mm and 2.9 ± 1.2 mm for the acetabular cup and femoral head respectively. This FCOR visualized in AR could be used by the surgeon as a guide for femoral neck cut, broaching and prosthetic stem insertion. On the acetabular side, this AR visualization could help guide the reaming depth and cup insertion [62].

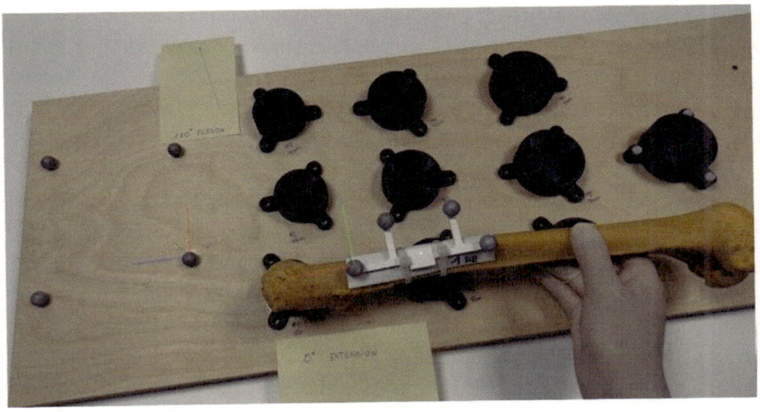

Figure 5.
Hip phantom consisting of dried human cadaveric femurs and 3D printed acetabular cup analogues. Both the femurs and acetabular cups are equipped with an IR reflective tracker/marker.

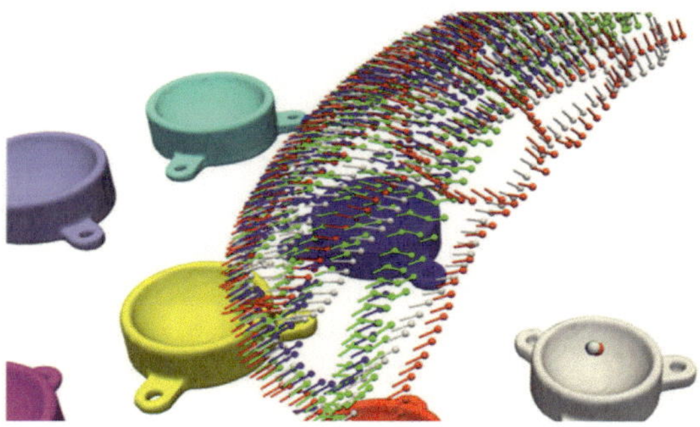

Figure 6.
Point clouds generated by the IR tracking software based on the displacement of the femoral tracker relative to the acetabular tracker. Two observers rotated each femur twice in its matching cup, producing 4 measurements per hip joint.

5. Future perspectives for AR in neurosurgery and orthopedic surgery

AR-based navigation is a hot topic in scientific literature with a rapidly growing number of new systems and use cases. However, the application of AR-based navigation in neurosurgery and orthopedic surgery is still in its infancy. As such, it will require further refinement and validation before it could be widely adopted during routine surgical procedures. For AR-based navigation to break through in clinical practice, some important issues should be addressed. First, user-friendliness plays an important role. The AR-based navigation must be efficiently embedded into the existing well-established surgical workflow of the procedure at hand. Secondly, specific attention must be paid to the desired image-to-patient registration and tracking technologies for each use case. Ideally, we will evolve from the periodic rigid registration techniques used today, to multisensory automatic markerless non-rigid registration techniques. In this way, the information from multiple sensors could be combined and deformation of the patient's anatomy could be taken-into-account without dead angles, warranting that AR guidance remains accurate throughout the entire surgical intervention. Thirdly, we should focus on the development of dedicated surgical headsets with short working distances and optimal visualizations without delays to reduce visual fatigue. In addition, dynamic user interfaces should be developed that enable users to activate suitable AR information on demand in different phases of a workflow and minimize the interference of AR information with the perception of real surgical situations [63].

In our opinion, the introduction of artificial intelligence and deep learning methods could facilitate the implementation of these suggestions and facilitate the evolution from conventional navigation systems towards AR navigation. The development of smart systems that select the most suitable data from a multi-sensor tracking stream and adjust the registration and navigation in real time would represent a groundbreaking improvement. In a second step, these smart systems could also follow the surgical procedures they are guiding and automatically display and adapt the desired navigation information. Finally, smart AR-based navigation could be combined with sensing technologies to provide direct and personalized feedback to surgical tools when, for example, the surgeon might commit a critical error. Overall, with further technological innovations and clinical validation of the systems, AR-based surgical navigation has the potential to become an indispensable, timesaving, risk-reducing and accuracy-improving technology in surgery.

6. Conclusion

In this chapter, we highlighted different AR technologies currently in development for neurosurgery and orthopedic surgery. Further in vitro and in vivo studies followed by prospective randomized clinical trials are needed to refine and assess efficacy of these systems in daily practice. However, we believe that at the current rate of technological development, AR-based navigation can become an indispensable part of surgical navigation within a few years.

Conflict of interest

The authors declare no conflict of interest.

Author details

Quentin Neuville[1,2]*, Thierry Scheerlinck[3,4] and Johnny Duerinck[1,2]

1 Department of Neurosurgery, Vrije Universiteit Brussel, Universitair Ziekenhuis Brussel, Brussels, Belgium

2 Research Group Center for Neurosciences (C4N-NEUR), Vrije Universiteit Brussel, Brussels, Belgium

3 Department of Orthopedic Surgery and Traumatology, Vrije Universiteit Brussel, Universitair Ziekenhuis Brussel, Brussels, Belgium

4 Research Group Beeldvorming en Fysische wetenschappen (BEFY-ORTHO), Vrije Universiteit Brussel, Brussels, Belgium

*Address all correspondence to: quentin.neuville@uzbrussel.be

References

[1] Härtl R, Lam KS, Wang J, Korge A, Kandziora F, Audigé L. Worldwide survey on the use of navigation in spine surgery. World Neurosurgery. 2013;**79**(1):162-172

[2] Ivan ME et al. Augmented reality head-mounted display-based incision planning in cranial neurosurgery: A prospective pilot study. Neurosurgical Focus. 2021;**51**(2):E3

[3] Krishnan R, Hermann E, Wolff R, Zimmermann M, Seifert V, Raabe A. Automated fiducial marker detection for patient registration in image-guided neurosurgery. Computer Aided Surgery. 2003;**8**(1):17-23

[4] Léger É, Drouin S, Collins DL, Popa T, Kersten-Oertel M. Quantifying attention shifts in augmented reality image-guided neurosurgery. Healthcare Technology Letters. 2017;**4**(5):188-192

[5] Liu A et al. Clinical accuracy and initial experience with augmented reality-assisted pedicle screw placement: The first 205 screws. Journal of Neurosurgery. Spine. 2021;**2021**:1-7

[6] Sorriento A et al. Optical and electromagnetic tracking Systems for Biomedical Applications: A critical review on potentialities and limitations. IEEE Reviews in Biomedical Engineering. 2020;**13**:212-232

[7] Hersh A et al. Augmented reality in spine surgery: A narrative review. HSS Journal. 2021;**17**(3):351-358

[8] Molina CA, Sciubba DM, Greenberg JK, Khan M, Witham T. Clinical accuracy, technical precision, and workflow of the first in human use of an augmented-reality head-mounted display stereotactic navigation system for spine surgery. Operation Neurosurgery (Hagerstown). 2021;**20**(3):300-309

[9] Molina CA et al. A cadaveric precision and accuracy analysis of augmented reality-mediated percutaneous pedicle implant insertion. Journal of Neurosurgery. Spine. 2020;**34**(2):316-324

[10] Luebbers H-T et al. Comparison of different registration methods for surgical navigation in cranio-maxillofacial surgery. Journal of Cranio-Maxillo-Facial Surgery. 2008;**36**(2):109-116

[11] Hong J, Hashizume M. An effective point-based registration tool for surgical navigation. Surgical Endoscopy. 2010;**24**(4):944-948

[12] Kriechling P, Loucas R, Loucas M, Casari F, Fürnstahl P, Wieser K. Augmented reality through head-mounted display for navigation of baseplate component placement in reverse total shoulder arthroplasty: A cadaveric study. Archives of Orthopaedic and Trauma Surgery. 2023;**143**(1):169-175

[13] Chen X et al. Development of a surgical navigation system based on augmented reality using an optical see-through head-mounted display. Journal of Biomedical Informatics. 2015;**55**:124-131

[14] Wu J-R, Wang M-L, Liu K-C, Hu M-H, Lee P-Y. Real-time advanced spinal surgery via visible patient model and augmented reality system. Computer Methods and Programs in Biomedicine. 2014;**113**(3):869-881

[15] Ma L, Huang T, Wang J, Liao H. Visualization, registration and tracking techniques for augmented reality guided

surgery: A review. Physical Medical Biology. 2023;**68**(4)

[16] Li R, Si W, Liao X, Wang Q, Klein R, Heng P-A. Mixed reality based respiratory liver tumor puncture navigation. Computational Visual Media. 2019;**5**(4):363-374

[17] Nguyen NQ et al. An augmented reality system characterization of placement accuracy in neurosurgery. Journal of Clinical Neuroscience. 2020;**72**:392-396

[18] Van Gestel F et al. Neuro-oncological augmented reality planning for intracranial tumor resection. Frontiers in Neurology. 2023;**14**:1104571

[19] Frantz T, Jansen B, Duerinck J, Vandemeulebroucke J. Augmenting Microsoft's HoloLens with vuforia tracking for neuronavigation. Healthcare Technology Letters. 2018;**5**(5):221-225

[20] Gibby JT, Swenson SA, Cvetko S, Rao R, Javan R. Head-mounted display augmented reality to guide pedicle screw placement utilizing computed tomography. International Journal of Computer Assisted Radiology and Surgery. 2019;**14**(3):525-535

[21] Alp MS, Dujovny M, Misra M, Charbel FT, Ausman JI. Head registration techniques for image-guided surgery. Neurological Research. 1998;**20**(1):31-37

[22] Widmann G, Stoffner R, Sieb M, Bale R. Target registration and target positioning errors in computer-assisted neurosurgery: Proposal for a standardized reporting of error assessment. International Journal of Medical Robotics. 2009;**5**(4):355-365

[23] Vassallo R, Rankin A, Chen ECS, Peters TM. Hologram stability evaluation for Microsoft HoloLens. In: Proceedings SPIE 10136, Medical Imaging 2017: Image Perception, Observer Performance, and Technology Assessment. 10 March 2017;**10136**. DOI: 10.1117/12.2255831

[24] Pietruski P et al. Supporting fibula free flap harvest with augmented reality: A proof-of-concept study. Laryngoscope. 2020;**130**(5):1173-1179

[25] Meulstee JW et al. Toward holographic-guided surgery. Surgical Innovation. 2019;**26**(1):86-94

[26] Van Gestel F et al. The effect of augmented reality on the accuracy and learning curve of external ventricular drain placement. Neurosurgical Focus. 2021;**51**(2):E8

[27] Skyrman S et al. Augmented reality navigation for cranial biopsy and external ventricular drain insertion. Neurosurgical Focus. 2021;**51**(2):E7

[28] Liebmann F et al. Pedicle screw navigation using surface digitization on the Microsoft HoloLens. International Journal of Computer Assisted Radiology and Surgery. 2019;**14**(7):1157-1165

[29] Farshad M, Fürnstahl P, Spirig JM. First in man in-situ augmented reality pedicle screw navigation. The Spine Journal. 2021;**6**:100065

[30] Farshad M et al. Operator independent reliability of direct augmented reality navigated pedicle screw placement and rod bending. The Spine Journal. 2021;**8**:100084

[31] Felix B et al. Augmented reality spine surgery navigation: Increasing pedicle screw insertion accuracy for both open and minimally invasive spine surgeries. Spine. 2022;**47**(12):865-872

[32] Abe Y et al. A novel 3D guidance system using augmented reality for percutaneous vertebroplasty: Technical note. Journal of Neurosurgery. Spine. 2013;**19**(4):492-501

[33] Viehöfer AF et al. Augmented reality guided osteotomy in hallux Valgus correction. BMC Musculoskeletal Disorders. 2020;**21**(1):438

[34] Londei R et al. Intra-operative augmented reality in distal locking. International Journal of Computer Assisted Radiology and Surgery. 2015;**10**(9):1395-1403

[35] Cho HS et al. Augmented reality in bone tumour resection: An experimental study. Bone Joint Research. 2017;**6**(3):137-143

[36] Hiranaka T et al. Augmented reality: The use of the PicoLinker smart glasses improves wire insertion under fluoroscopy. World Journal of Orthopedics. 2017;**8**(12):891-894

[37] Andress S et al. On-the-fly augmented reality for orthopedic surgery using a multimodal fiducial. Journal of Medical Imaging (Bellingham). 2018;**5**(2):021209

[38] Molina CA, Dibble CF, Lo S-FL, Witham T, Sciubba DM. Augmented reality-mediated stereotactic navigation for execution of en bloc lumbar spondylectomy osteotomies. Journal of Neurosurgery. Spine. 2021;**2021**:1-6

[39] Elmi-Terander A et al. Augmented reality navigation with intraoperative 3D imaging vs fluoroscopy-assisted free-hand surgery for spine fixation surgery: A matched-control study comparing accuracy. Scientific Reports. 2020;**10**(1):707

[40] Elmi-Terander A et al. Pedicle screw placement using augmented reality surgical navigation with intraoperative 3D imaging: A first In-human prospective cohort study. Spine. 2019;**44**(7):517-525

[41] Dennler C et al. Augmented reality in the operating room: A clinical feasibility study. BMC Musculoskeletal Disorders. 2021;**22**(1):451

[42] Tsukada S, Ogawa H, Hirasawa N, Nishino M, Aoyama H, Kurosaka K. Augmented reality- vs accelerometer-based portable navigation system to improve the accuracy of acetabular cup placement during total hip arthroplasty in the lateral decubitus position. The Journal of Arthroplasty. 2022;**37**(3):488-494

[43] Tsukada S, Ogawa H, Nishino M, Kurosaka K, Hirasawa N. Augmented reality-assisted femoral bone resection in total knee arthroplasty. JB JS Open Access [Internet]. 23 Jul 2021;**6**(3). DOI: 10.2106/JBJS.OA.21.00001

[44] Kurosaka K, Ogawa H, Hirasawa N, Saito M, Nakayama T, Tsukada S. Does augmented reality-based portable navigation improve the accuracy of cup placement in THA compared with accelerometer-based portable navigation? A randomized controlled trial. Clinical Orthopaedics and Related Research. 2023;**481**(8):1515-1523

[45] Kiarostami P et al. Augmented reality-guided periacetabular osteotomy-proof of concept. Journal of Orthopaedic Surgery and Research. 2020;**15**(1):540

[46] Chytas D, Malahias M-A, Nikolaou VS. Augmented reality in Orthopedics: Current state and future directions. Frontier in Surgery. 2019;**6**:38

[47] Unberath M et al. Augmented reality-based feedback for technician-in-the-loop C-arm repositioning. Healthcare Technology Letters. 2018;**5**(5):143-147

[48] Hu M-H, Chiang C-C, Wang M-L, Wu N-Y, Lee P-Y. Clinical feasibility of the augmented reality computer-assisted spine surgery system for percutaneous

vertebroplasty. European Spine Journal. 2020;**29**(7):1590-1596

[49] Iop A, El-Hajj VG, Gharios M, de Giorgio A, Monetti FM, Edström E, et al. Extended reality in neurosurgical education: A systematic review. Sensors [Internet]. 2022;**22**(16):6067. DOI: 10.3390/s22166067

[50] Logishetty K, Western L, Morgan R, Iranpour F, Cobb JP, Auvinet E. Can an augmented reality headset improve accuracy of acetabular cup orientation in simulated THA? A randomized trial. Clinical Orthopaedics and Related Research. 2019;**477**(5):1190-1199

[51] Ha J et al. Opportunities and challenges of using augmented reality and heads-up display in orthopaedic surgery: A narrative review. Journal of Clinical Orthopedic Trauma. 2021;**18**:209-215

[52] Condino S, Carbone M, Piazza R, Ferrari M, Ferrari V. Perceptual limits of optical see-through visors for augmented reality guidance of manual tasks. IEEE Transactions on Biomedical Engineering. 2020;**67**(2):411-419

[53] Ferrari V, Carbone M, Condino S, Cutolo F. Are augmented reality headsets in surgery a dead end? Expert Review of Medical Devices. 2019;**16**(12):999-1001

[54] Fischer M et al. Preclinical usability study of multiple augmented reality concepts for K-wire placement. International Journal of Computer Assisted Radiology and Surgery. 2016;**11**(6):1007-1014

[55] Huyette DR, Turnbow BJ, Kaufman C, Vaslow DF, Whiting BB, Oh MY. Accuracy of the freehand pass technique for ventriculostomy catheter placement: Retrospective assessment using computed tomography scans. Journal of Neurosurgery. 2008;**108**(1):88-91

[56] Scheerlinck T. Cup positioning in total hip arthroplasty. Acta Orthopaedica Belgica. 2014;**80**(3):336-347

[57] Scheerlinck T. Primary hip arthroplasty templating on standard radiographs. A stepwise approach. Acta Orthopaedica Belgica. 2010;**76**(4):432-442

[58] Liebs TR, Nasser L, Herzberg W, Rüther W, Hassenpflug J. The influence of femoral offset on health-related quality of life after total hip replacement. Bone Joint Journal. 2014;**96-B**(1):36-42

[59] Jolles BM, Zangger P, Leyvraz P-F. Factors predisposing to dislocation after primary total hip arthroplasty: A multivariate analysis. The Journal of Arthroplasty. 2002;**17**(3):282-288

[60] Konyves A, Bannister GC. The importance of leg length discrepancy after total hip arthroplasty. Journal of Bone and Joint Surgery. British Volume (London). 2005;**87**(2):155-157

[61] Renkawitz T et al. Leg length and offset differences above 5mm after total hip arthroplasty are associated with altered gait kinematics. Gait & Posture. 2016;**49**:196-201

[62] CARS 2023-Computer Assisted Radiology and Surgery. Proceedings of the 37th International Congress and Exhibition Munich, Germany, June 20-23, 2023. International Journal of Computer Assisted and Radiological Surgery. 2023;**18**(1):1-123

[63] Katić D et al. A system for context-aware intraoperative augmented reality in dental implant surgery. International Journal of Computer Assisted Radiology and Surgery. 2015;**10**(1):101-108

Chapter 10

The 3D Operating Room with Unlimited Perspective Change and Remote Support

Klaudia Proniewska, Damian Dolega-Dolegowski, Radek Kolecki, Magdalena Osial and Agnieszka Pregowska

Abstract

Information and communication technologies combined with extended reality improve diagnostics, medical treatment, and surgical operations courses. Thus, the new generation of devices, which enable displaying of holographic objects, allows visualizing a patient's internal organs, both anatomical and pathological structures, as interactive 3D objects, based on retrospective 2D images, namely computer tomography (CT) or magnetic resonance imaging (MRI). Remote users can observe ongoing surgery with additional features like standing in front of the operation table, walking around in, and changing the user's perspective without disrupting the operating doctors. Moreover, the operation can be recorded, then played with additional functionalities—remote pointing and painting, which is important in medical education. The operating surgeon may also ask remotely more experienced operators for consultation, and as a consequence, the number of complications and failed procedures can decrease. It can be also applied to robot surgeries as a first step to remote surgery.

Keywords: 3D operating room, extended reality, computer-integrated surgery, image-guided surgery, medical education

1. Introduction

Extended reality (XR) includes virtual reality (VR), augmented reality (AR), and mixed reality (MR) [1, 2]. It incorporates the spectrum of technologies, which allow you to combine and/or mirror the real world (i.e. the physical world) with the "digital twin world" providing the possibility of iterating among others. The computer-generated images and objects are presented in front of the user's eyes with head-mounted displays (HMDs), which provide a hand-free view of virtual objects, like text and images [3]. Most of the commercially available HMDs used optical see-through (OST) to enable the users to perceive the world through the set of optical components. Usually displays contain optical components like half mirrors, birdbaths, free-form prism, and optical waveguides. Since XR technology enables the superimposition of two-dimensional (2D) and three-dimensional (3D) objects, it has recently been applied in various fields of life, in particular in medicine, with special emphasis on all types of surgery, including image-guided surgery (IGS) and computer-integrated surgery (CIS), while it has the potential to improve the safety and efficiency of medical procedures [4, 5]. More efficiently planned surgical interventions may result in shorter recovery times and better treatment outcomes. Thus, the transmission to less invasive surgery requires the development of visualization techniques without limited perspective.

The XR-based solution can help surgeons integrate 2D images obtained from ultrasound, magnetic resonance imaging (MRI), or computed tomography (CT) in the DICOM format (digital imaging and communication data in medicine) with the 3D operation view through overlay virtual objects over a surgical field. It enables the enrichment of the operating field with computer-generated digital images, especially in the visualization of tumors and anatomical structures [6]. And in this way, the application of the 3D anatomical information in preoperative planning, and creates the possibility of integrating the preoperative model with the intra-operative scenario and guides the surgeon in real time [7–9]. The advantage of systems based on XR technology in comparison to traditional auxiliary displays is the fact that the surgeons do not have to change their line of sight between the operating scene and the auxiliary display, which significantly reduces the operating time. Another XR application possibility in surgery is connected with the support or even replacement of the surgeon apprenticeship by high-fidelity surgical simulators [10]. However, implementation in the surgeon's workflow is hampered by a lack of clinically useful application development requirements.

Here, we proposed a 3D operating room with unlimited perspective change and remote support, which is based on Microsoft HDMs—HoloLens 2, and Intel RealSense cameras. The basic idea of the system is shown in **Figure 1**. The system allows the users (surgeons and/or medical students) to interact with the headset using a verbal command or a simple gesture such as hand movements or eye movements. The proposed approach is completely sterile and can be successfully implemented in clinical procedures. It also enables the operating surgeon to share his field of vision (headset vision) with other users (surgeons or medical students), which can be located in any place without disturbing the workflow. Moreover, it is possible to change the position, and angle of the medical procedure observation as well as subsequent playback and analysis of the recorded procedure from different perspectives, which is of great importance in the training of future medical staff. We also overview available XR-based solutions in surgery, and we show the lines of XR technology development in the field of surgery.

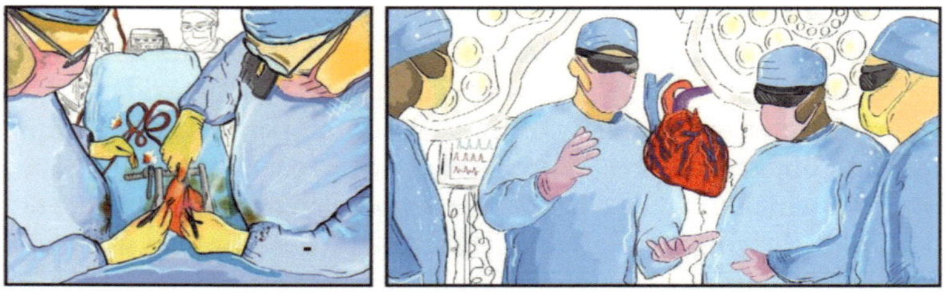

Figure 1.
The scheme of XR-based support system for surgeons.

DOI: http://dx.doi.org/10.5772/intechopen.1002252

2. Extended reality-based technologies in surgery

In recent years, surgery has undergone significant technological advancement, especially in the field of the application of computer vision, like image-guided surgery and computer-integrated surgery) [11]. The development of extended reality allows surgeons to visualize medical data to assist in the execution of both complex and routine medical procedures. It enables the incorporation of the operation field and anatomical landmarks with navigation guidance [12]. Thus, the XR-based solution may contribute to the improvement of surgical procedures from the point of a surgeon's view in comparison to the classical one. When it comes to XR, surgery can be divided into two application areas, namely, preoperative planning, and intraoperative support [4]. However, it is required to be proven to be safe and reliable in order to be applied in the operating rooms.

2.1 Preoperative planning

The crucial part of all types of surgery is preoperative planning. The factors like medical history, surgical past, possible comorbidities, medications, and physical conditions must be taken into account [13]. Since extended reality, which has a wide spectrum, which is oscillating through virtual, augmented, and mixed realities enables 3D visualization, can be used for surgery preoperative planning to provide insight into the subject's anatomy. Thus, the construction of the 3D preoperative model required the processing of the CT or/and MRI scan images. Next, these scan images are segmented, mostly manually, but also automatically, for example, using algorithms based on artificial intelligence (AI) [14, 15]. On their basis, a three-dimensional image of the organ and its abnormalities, arteries, veins, and other anatomical structures is developed and then rendered. The final phase is exporting the 3D model to HDMs.

Initial results from the use of XR have been reported for virtually every branch of surgery, in particular, preoperative 3D visualization is profitable in plastic and craniomaxillofacial surgery, where the result very often determines the patient's decision about surgery [16, 17]. VR-based [18, 19] solutions (PulmoVR) help to correct the lung-tumor placement in 52.00% of cases, and even in 10.00% of cases contributed to the patient's preservation of the lung despite prior indications for lung removal. PulmoVR creates the lung digital twin based on CT-scan images in front of the surgeries eyes. The structures like airways, arteries, veins, and lungs segment are visualized. The system enables the visualization of those parts of the patient's lungs that the doctor is interested in. Another XR-based solution, in opposition to HMDs, that uses 3D displays as 3D TVs is Echopixel, which allows visualization of arteries in patients with pulmonary atresia.

Other issues of the XR application in the medical sector are connected with the subjects-doctor communications. It is especially important in preoperative planning, while it enables better understanding by the patient of the future treatment (the essence of the operation and its course) and possible complications that may occur after the procedure, which affects the patient's trust in the doctor. Many procedures do not end only with the resection of the diseased tissue, very often it also requires pathological and psychological rehabilitation. Raising awareness is also crucial for the future effectiveness of targeted therapy, and here XR-based technologies give a huge opportunity [20].

2.2 Intraoperative support

Trauma surgeons are often faced with complex situations that keep track of X-ray procedures. Improved 3D imaging would provide both, an increase in the efficiency of the medical procedure, and reduce the time of exposure to X-rays of the medical team [12, 21]. Thus, the extended reality has a huge potential to become an intraoperative support for surgeons by visualization and superimposition of CT or MRI image scans collected before surgery on the patient, thus improving navigation during surgery [22]. The intraoperatively 3D model/objects of the desired anatomical structures can be displayed in HDMs [23]. To visualize the surgeon's field of view, a camera is needed to accurately monitor the distance and angle of the object of interest. Some proposed solutions also include accelerometers, optical sensors, GPS, and gyroscopes as well as navigation systems [24]. One can say that XR-based technologies are the interface between surgeons and computer-generated objects, which are based on the medical documentation of subjects.

Intraoperative navigation became an essential part of complex medical procedures. This navigation can use 3D objects obtained from previous medical records on the part of the patient of interest during the medical procedure. For example, in ref. [25], the augmented reality-assisted navigation system (ARAN) was shown. It enables real object positioning according to the information contained in the CT image scans. It turned out that XR-based approaches enable obtaining better clinical output in knee arthroplasty. XR systems can also enable effective navigation during more demanding operations, including heart and brain surgeries while information concerning spatial comprehension and often the precise localization of structures (like deep-seated tumors) is required [26–28]. The AR-supported surgical navigation platform DEX-Ray was proposed in ref. [26]. It enables the visualization of the tumor and surrounding structures and the anatomy of the venous, including critical draining veins. The system also assisted in determining the optimal scalp flap and bone window for surgical access by selection transparency of the scalp and bones. It was tested during the resection of meningiomas in the falcine, convexity, and parasagittal regions [29]. In ref. [30], the DEX-Ray system was combined with mixed reality deceive—Microsoft HoloLens.

Moreover, minimally invasive surgery (MIS) technology has been developed to limit access to wound injuries and reduce the incidence of postoperative complications [31]. An interesting proposition is also to combine the XR technologies with robotic-assisted surgery (RAS) [32, 33]. In this case, the XR is responsible for the user interfaces, which allow for the reduction of the phenomenon of looking away and increases the situational awareness of the user. Another example is application of the augmented reality to enhance endoscopic video during *in-vivo* robot-assisted radical prostatectomy (RARP) [34]. The solution enables putting a virtual 3D object of the patient's prostate on top of its 2D counterpart in real-time.

2.3 Medical education

Extended reality is also increasingly used in the field of education, especially when it comes to medical education. For example, the Stanford Virtual Heart Project, in which doctors use VR for visualization and understanding of congenital heart defects [The Stanford Virtual Heart—Stanford Medicine Children's Health (stanfordchildrens.org)]. It visualizes the normal and abnormal anatomy of the heart. First, it was aimed at families of children with heart defects but later spread to students as well. It allows users to see the anatomy of the heart, and the blood flow inside it and observed how the defect interferes with the proper functioning of the organ. Virtual heart is also used to visualize the medical procedures conducted by pediatric heart surgeons to repair the not correctly functioning heart. Another solution is developed by Case Western Reserve University and Cleveland Clinic, a HoloAnatomy application tailored to run on Microsoft HoloLens devices to learn human anatomy (https://engineering.case.edu/HoloAnatomy-honors). The effectiveness of HoloAnatomy was analyzed during medical students' courses [35, 36] [HoloAnatomy® and MR in the Jagiellonian University Medical College (JUMC) https://mrame.cm-uj.krakow.pl/], see **Figure 2**. HoloAnatomy® Suite takes advantage of these technical benefits and provides students with an opportunity to learn anatomy in a completely new way. The heart and soul of every HoloAnatomy® lesson is a holographic slideshow. It consists of accurate, three-dimensional models with a controlled number of anatomical structures. There is also a possibility to add other educational materials to the slideshow. What is important is the teacher and students see the model in the same place. This enables the tutor to point or magnify any structure from the model he/she wants to show. All things considered, HoloAnatomy® not only allows students to better understand anatomy, especially topographical relations between anatomical structures but also maintains an ability to communicate with the tutor and interact with the model. This course aimed to prepare JUMC students to use the latest diagnostic technology, the so-called extended reality, thanks to which they will achieve unique competencies valued in the labor market. In addition, the aim was to improve competencies and professional qualifications through participation in specialized training and study visits tailored to the needs of students. Training and visits are addressed to JUMC students. During the semester program, students will gain new competencies thanks to the synergy of several activities:

1. Specialized training in medical data visualization methods using XR technology;

2. Access to audiovisual materials in the field of techniques and methods of using XR (e.g., in cardiology, neurology, etc.);

3. Participating in study visits to employers using pictorial tools for visualizing medical data, e.g., XR.

Students qualified for the program gained access to the latest technology (software and hardware). The training was conducted by highly qualified staff with experience in both medical diagnostics and augmented reality technology used in medicine. The JUMC didactic staff have developed syllabuses and training scenarios, as well as will evaluate the competencies of students qualified for the project. Thanks to the availability of audiovisual materials, students could additionally improve their skills. Study visits of students to employers (hospitals, diagnostic imaging laboratories, etc.) located in Cracow, Poland were an integral part of the program. Participation of students in the project will improve their competencies and professional qualifications (gaining new knowledge and skills and using it in practice), group work, and problem-solving skills (group cooperation, data analysis).

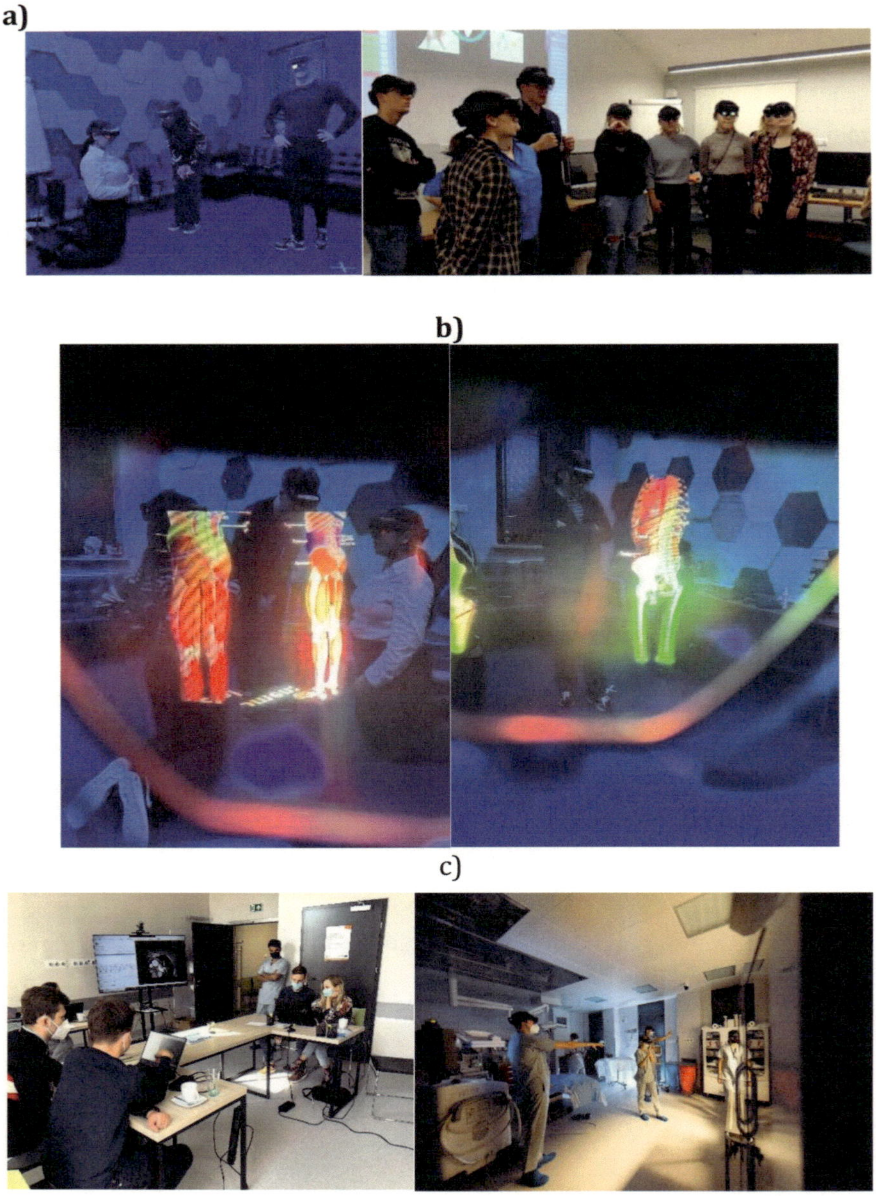

Figure 2.
XR support in medical courses, (a) the view person without HDMs, (b) the view of the HDMs users, and (c) conducting study visits for students of Collegium Medicum of the Jagiellonian University. Students have the opportunity to familiarize themselves with the visualization of medical data using extended reality—advanced imaging as a tool used to optimize pretreatment planning and intraprocedural monitoring in clinical environments.

The preoperative virtual exercises are important for the effective training of medical staff. They allow you to shorten the training time, increase the level of its results and reduce costs. XR not only allows you to visualize anatomical structures in a more realistic and accessible way but also engages the student more than books and autopsies [37]. Thanks to the XR application, the users can undergo training in a time frame adapted to their needs, without stress related to the possibility of making a mistake resulting in permanent damage to patients' health or even their death. In addition, without restrictions related to the subjects' time under anesthesia, consumption of blood and plasma and other substances necessary during the operation, costs of the operating room, and participation of other staff. The main XR advantage is the ability to make mistakes and correct them with helpful hints. An interesting proposal is also the implementation of XR technology for medical exercises of deep space mission crews who have to deal with various health disorders in space [38].

Long-term comparative observations were also carried out between medical students using XR technology for learning and students learning traditionally [39]. It turned out that the first group of students was characterized by greater determination in pursuing a career and better grades. In ref. [40], it was shown that XR-based technologies have a positive influence on medical students by the increase topic interest, focus, and motivation.

2.4 Relation with patients: mutual understanding and anxiety management

However, not only doctors, future doctors, and other medical staff can benefit from the implementation of XR technology in practice. Sometimes it is very difficult to find a common ground of communication between the patient and the doctor [41]. Attractive and, as far as possible, real visualization of a medical procedure can effectively move mutual communication to a higher level of understanding. Thus, XR can make it easier for the patient to assimilate and understand the information provided by the doctor regarding the disease itself and the process of its treatment. It contributes to anxiety reduction, which is connected with the medical procedure in all stages, namely, preoperative, intraoperative, and postoperative [42, 43]. XR has been shown to reduce stress in patients but does not affect objective measurements of patients' physiological status [44].

3. Three-dimensional operation room

In this paper, the XR-based support sterile system for surgeons in the operating rooms was designed and implemented. The proposed solution is presented in **Figures 1** and **3**. The operating surgeon may share his field of headset vision with other users, which may be located. Users can change the position, and angle of the medical procedure observation as well as subsequent playback and analysis of the recorded procedure from different perspectives, without impact on the work of the operating surgeon presented in **Figure 4**. The Intel RealSense cameras were applied to provide XR streaming from 3D cameras to HoloLens 2 glasses.

The XR streaming, which is shown in **Figures 3–5,** is carried out as follows:

Step 1. *The gathering of 3D data by Intel RealSense cameras*. The cameras required PC connections to reduce the size of the system and mobility issues (fewer cables due to the requirement only of the source of power like a power bank), in our case, the cameras are connected to a Nvidia Jetson microcomputer. When cameras catch data for 3D processing, they will pass it to the Nvidia device, which collects data and performs basic transition and smoothing for future processing. Data are locally stored at the connected SSD drive. When recording and processing of data are finished, data is sent to local Network Area Storage (NAS). Note, that one camera cannot

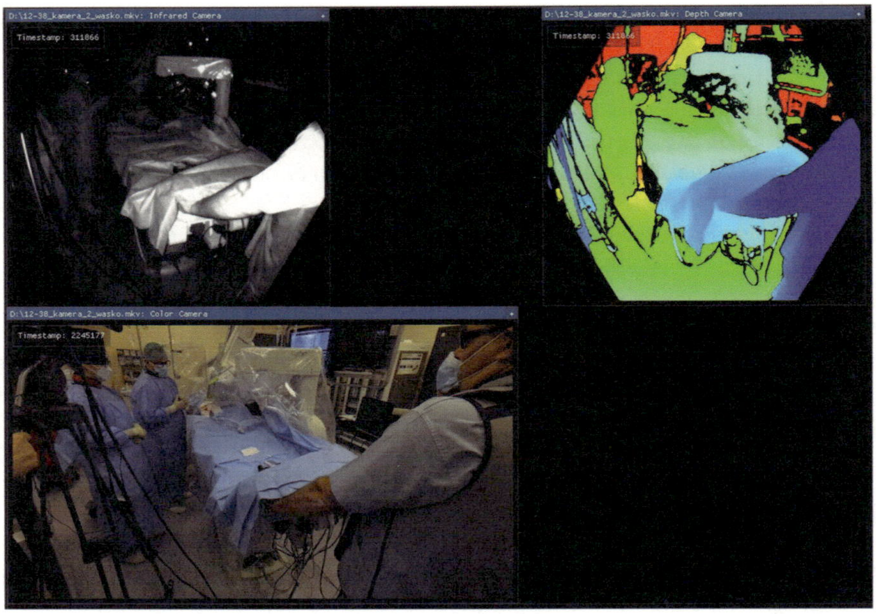

Figure 3.
A set-up environment for The 3D operating room with unlimited perspective change and remote support. The operating room is captured by a 3D camera divided into infrared, depth, and color cameras. Three-dimensional-captured images can be transferred to extended reality HDMs.

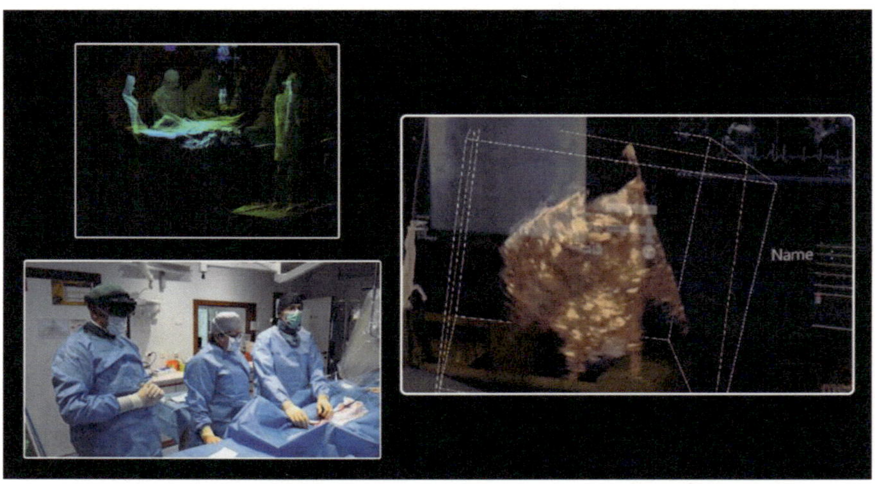

Figure 4.
Three-dimensional visualization in surgery involves the use of advanced imaging techniques and intraoperative imaging, to create detailed 3D representations of the surgical field e.g., 3D ultrasound with real-time data transfer to XR. Surgeons can then view these images on specialized displays or through extended reality headsets, enhancing their depth perception and understanding of anatomical structures.

cover the entire space/room. The application of the 2–3 pairs of cameras with Nvidia microcomputer is needed. Each of them will record data at the same time from a different position and place (different perspectives). As a result, few recordings will be achieved.

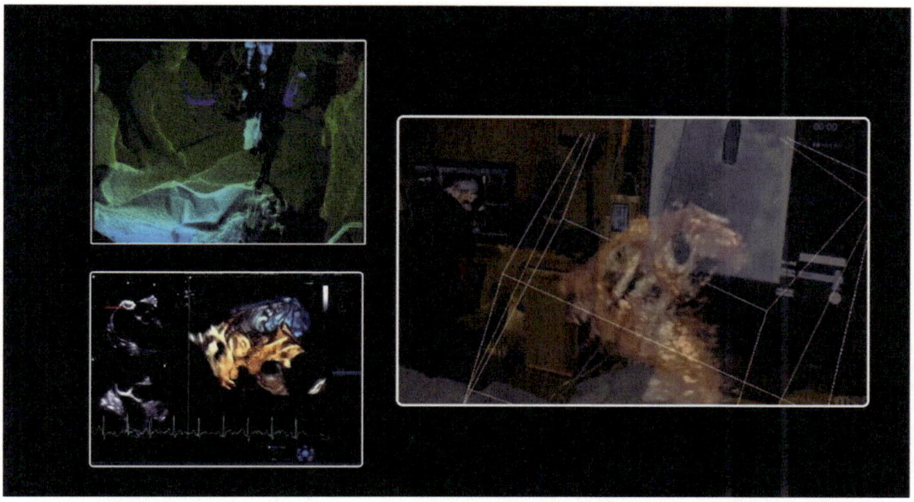

Figure 5.
In a traditional operating room, surgeons typically rely on two-dimensional displays to visualize the surgical site. However, with the advent of advanced imaging technologies and XR, 3D operating rooms have emerged to enhance surgical precision and improve patient outcomes.

Step 2. *Automatic process of combining recordings registered by Intel RealSense camera into one large 3D recording.* Recordings, which are registered in Step 1, will be combined into one recording. For this purpose, the computationally powerful PC, and the load stored on NAS data (recordings) will be combined in one 3D large-size recording of the considered scene.

Step 3. *Streaming in XR-based devices*. A separate application is needed. It will load created earlier recordings and render from it 3D view according to data in it. This will be next streamed to XR-based HDMs, in our case HoloLens 2. While a large number of data and limited performance of HDMs entire process of 3D image rendering needs to be done on an external PC and streamed over WiFi to HoloLens 2 glasses. At the same time while streaming HoloLens 2 glasses will be sent to these PC, the actual position of HDMs in space and moves to allow the computer to render and stream the next view for the users.

Thus, remote support in surgery involves the use of telemedicine technologies to enable remote collaboration and consultation. Surgeons can connect with experts or specialists who are not physically present in the operating room, allowing for real-time guidance, knowledge sharing, and collaboration, see **Figures 6–8**.

4. Extended reality-based application for surgeons

Designing the XR applications is fundamentally different from designing applications intended for flat screens of computers or tablets, while the user interaction is different. Extended reality has more control over processes and their design. The designed application must take into account the way the user communicates with HDMs. For example, HoloLens 2 glasses work with human senses, and human preferences vary from person to person. While HoloLens 2 enables gestures, eye movements, or voice commands control, this must be taken into account during the

Figure 6.
Adjustable perspective refers to the ability to change the viewing angle or vantage point within a surgical environment. This can be achieved through the use of original assisted surgery systems or advanced imaging technologies that allow surgeons to support the extended view of the surgical field.

Figure 7.
Remote 3D medical consultation refers to the practice of utilizing advanced technologies to provide medical consultations and expertise remotely, with the added benefit of three-dimensional visualization.

design process. Visual control is still underestimated and the combination of voice, hand gestures, and gazes can create incredibly fluid experiences with contextual menus appearing and disappearing as the user looks at something significant. This will become even truer when eye tracking becomes the standard in this area. A big challenge in XR-based application design is the 3D user interfaces, while developing a 3D graphical interface that engages the user visually and has emotional significance is an important part of the design process, especially in medical aspects. The designing and implementation of the XR-based application, which will be customized to the Microsoft HoloLens 2 glasses, to display the 3D large-size recording are as follows:

Figure 8.
Healthcare professionals can use video conferencing tools to communicate with healthcare providers remotely. In this case extended reality overlays virtual elements onto the real-world environment, providing an enhanced view of the medical data. In the context of remote consultations, XR can be used to display three-dimensional medical images, such as CT image scans or MRI data, directly onto the patient's body or relevant objects. This allows for a more accurate assessment of the medical condition and facilitates real-time discussions between healthcare professionals.

Step 1. *Segmentation of Dicom data* (**Figure 9**). Every CT or MRI image scans contain a lot of unneeded data, which slows down device performance during the display of the model. Due to that segmentation process is required to leave only important parts of these data. This also allows the creation of separate models for bones, tissue, or even entire organs. It can be achieved by for example the open-source 3Dslicer application (https://www.slicer.org/).

Step 2. *Validation and error removal* (**Figure 10**). Segmented and saved models also require validation, which can be done using 3D graphic tools like the open-source Blender application (https://www.blender.org/). The main purpose is to eliminate duplicated layers on models or broken parts of the surface. Such errors can stop the model from being correct display in the application.

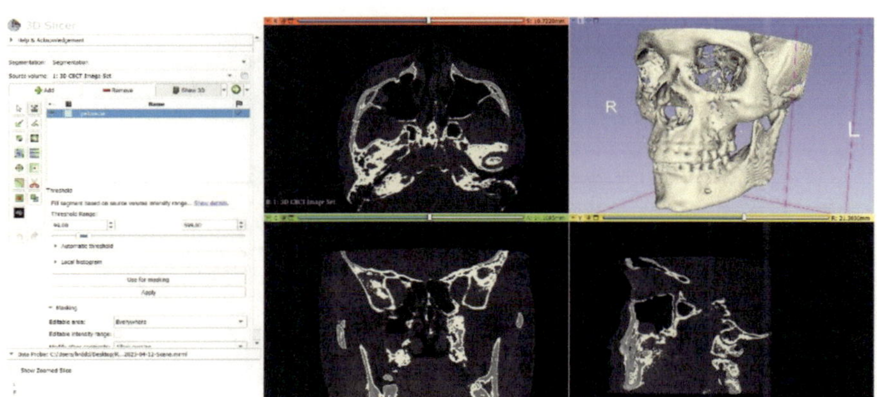

Figure 9.
The segmentation process is done in 3Dslicer.

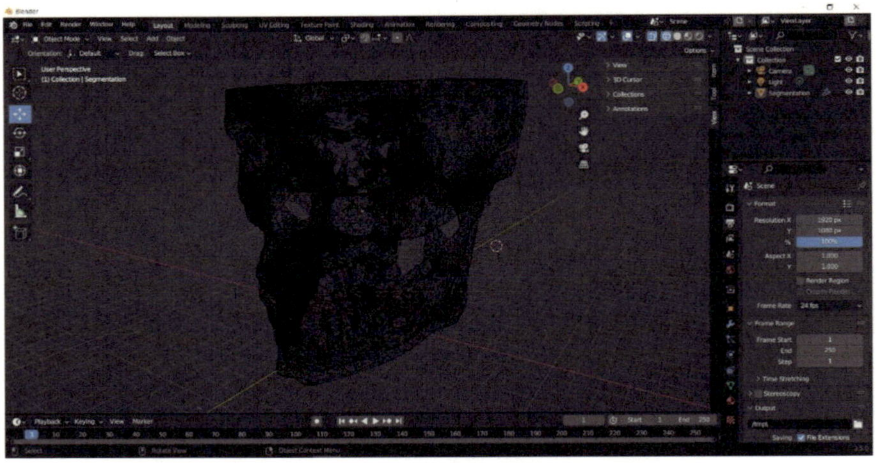

Figure 10.
Model validation using blender.

Step 3. *Application engine configuration* (**Figure 11**). In the case of HDMs like Microsoft HoloLens 2, the simplest way to create an application is to use a 3D creation tool (engine) such as Unreal Engine (https://www.unrealengine.com/en-US) or Unity (https://unity.com). Each of them requires a precise configuration allowing the application to get the best performance and quality. Several configuration settings can be set to improve the speed of the engine and the number of frames generated during the application run.

Step 4. *Importing models, assigning capabilities, and manipulation* (**Figure 12**). One of the last steps during the creation of a simple application is to import a segmented model and assign to it its capabilities. This means that displayed by the engine model will have actions and reactions. For example, to allow the model to be moved we need to activate a change of location on it. If the user wants to move and manipulate the position of the model it also requires to add to it reactions for hand gesture input. For example, if there will be a grab gesture, performed model should be moving with the user's hand until such gesture is stopped.

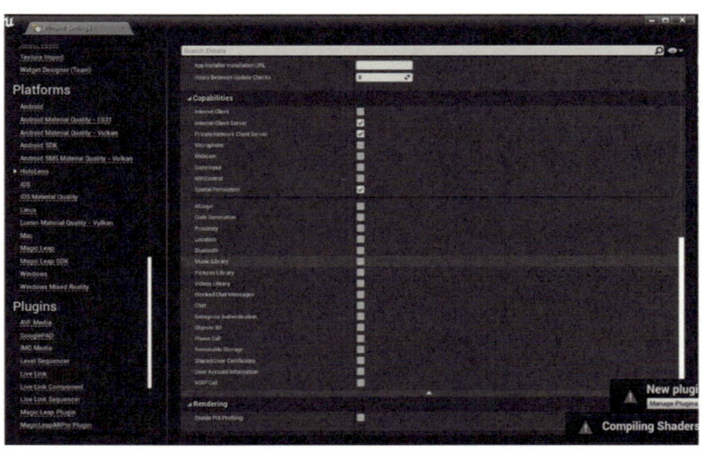

Figure 11.
Project configuration in unreal engine for Microsoft HoloLens 2 device.

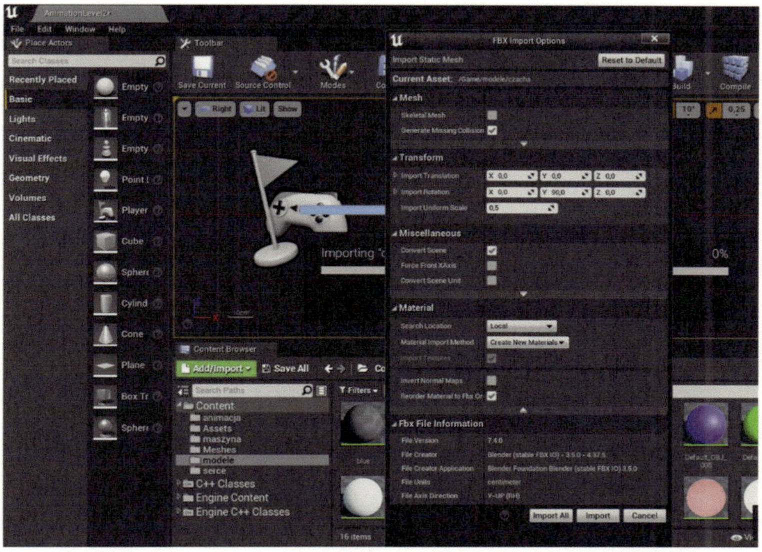

Figure 12.
Importing model and configuring its data.

Step 5. *Packaging application for the device*. The final part is to assign development certification, and application details and pack the application into a file that any Microsoft HoloLens 2 glasses will be able to simply and quickly use for installation. The process of packing application is very automated but includes the generation of all shadows and graphics of the model so it can take very long.

After performing the above steps finally, the installation file is ready and can be distributed and installed on any Microsoft HoloLens 2 device.

5. Case study description

The purpose of the presented case report is to establish a novel approach to preoperative patient experience and pre and intraoperative planning using extended reality 3D visualization of MRI or/and CT DICOM file segmentations. For patient experience, radiology can be complicated for patients to understand and fully appreciate, so we propose that seeing their 3D hologram, including segmented structures and highlighting the lesion would reduce stress levels, and increase relatability and personal engagement with treatment. Concerning pre and intraoperative utility, these segmentations visualize and differentiate between portal, venous, and arterial vasculature in one model, highlighting structures and the lesion all to simplify planning. Because the XR model is projected using see-through Microsoft HoloLens 2 goggles, the surgical team can wear them while seeing both the sterile field and hologram. Additionally, they can interact with the hologram without exiting or compromising the sterile field.

This XR-based hologram was created using a preoperative MRI image scan in 3D Slicer, an open-source platform for the analysis and visualization of medical images. After the segmentation process presented in **Figure 13**, which was validated by a radiologist on staff for accurately labeling the lesion and visible vessels, it was presented to the patient and surgical staff *via* the Microsoft HoloLens 2 the day before surgery. The objectives of this study were two-fold. First was proved that showing patients

a)

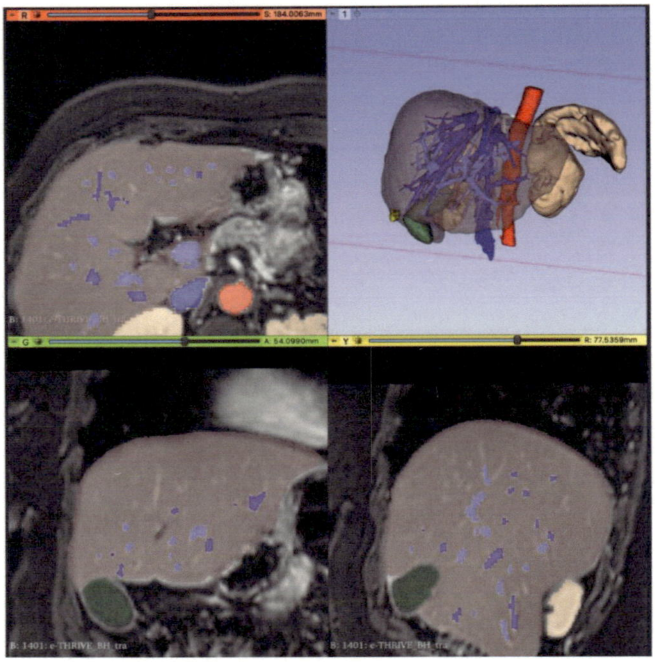

b)

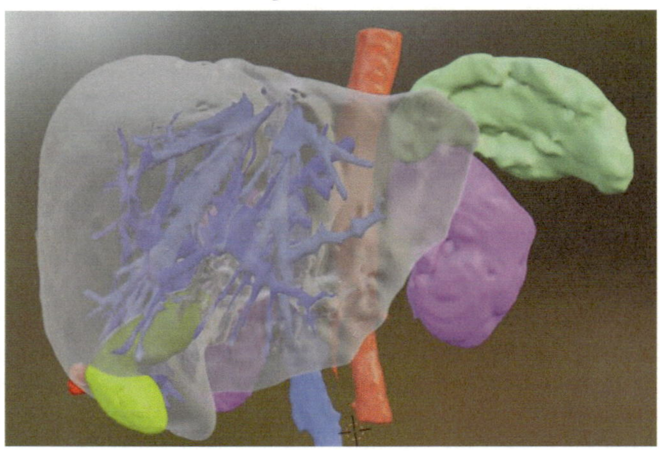

Figure 13.
The segmentation procedure (a) segmentation and volume rendering, including medical MRI image scan. The type of view was denoted by colors: Horizontal red bar, coronal green bar, and sagittal yellow bar, (b) magnified segmented object.

a 3D visualization of radiological scans provides unique benefits that help calm, inform, and engage with their care. Second, was sought to explore both the utility of using this extended reality visualization for preoperative planning and the practicality of being an intraoperative reference.

Medical staff reported that the proposed method of visualization reduced the mental effort required in of keeping track of structures and vasculature between

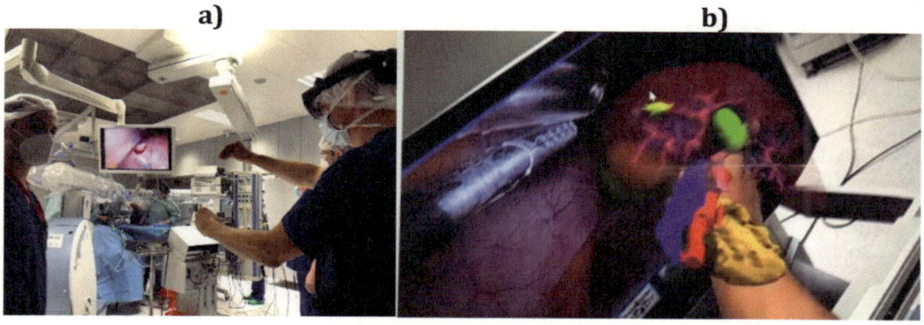

Figure 14.
System validation in the operating 3D room with the unobstructed field of view while wearing the Microsoft HoloLens 2 (a) view of the operator who is viewing the model (b) comparison of a real organ in the operating room and a digital organ displayed using XR-based device.

slices, demonstrating utility in the preoperative planning stage. They also commented that because the Microsoft HoloLens 2 goggles do not obstruct the field of view like other XR devices, this technique aids intraoperatively by serving as a reference point that can be readily accessed at any time without exiting or compromising the sterile field. An example of such a medical procedure is shown in **Figure 14**. Thus, the main challenge is the time required to complete a segmentation for visualization. The algorithms applied using 3D Slicer have been fully validated and automate much of the procedure, however, manual edits are required to complete a segmentation. The algorithms function by tracking voxel intensity between slices and creating a continuous extrapolation in 3D. However, there will be an associated mastery curve to reduce that time as the individual performing segmentations gain familiarity with 3D Slicer. On the other hand, the XR-based visualization of radiological scan segmentations using the Microsoft HoloLens 2 shows excellent promise in the field of surgery for both patient and operator experience. Patients can be more engaged with their care, have a greater understanding of their condition, and decreased preoperative anxiety. A surgeon can quickly and easily review a segmentation for surgical planning and later actively use it as a reference intraoperatively without leaving the sterile field.

6. Discussion and conclusion

The proposed XR-based solution enables the preview and registration of performed medical procedures without restrictions resulting from the camera settings and the recorded perspective, i.e. unlimited perspective. It provides the support of surgeons, which are performing a complex medical procedure by other specialists in a given field located in a different location and have much more experience and knowledge concerning the performed procedure. From the perspective of the surgeons' improvement process, it is a huge advantage, because they gain a chance to gain more experience and skills in procedures that they do not perform usually. Moreover, they gain access to consultations and support from specialists to whom they do not have access before due to the distance and location of a given medical facility (e.g. smaller cities away from large specialist hospitals by up to hundreds of kilometers). Surgeons who are not proficient in more complex procedures can count on the support of more experienced specialists who can help with an unexpected course of the procedure, e.g., when the operated lesion covers a larger area than originally assessed and the question arises about the best form of cutting/removing the lesion in this case. Many areas of medicine, including diagnostics of neurological and psychiatric diseases, rehabilitation after strokes, and advanced research on the human brain, will undoubtedly benefit from the introduction of XR-based solutions. Also, the patients receive benefits in the form of opportunities to take new treatment options.

An important limitation of the XR-based system application is changes that occur in the human eye as a result of the aging process, including age-related presbyopia, inability to focus on close objects, resulting in blurred vision, and natural aging hinders accommodation, while the majority of surgeons is middle age or older [45]. The last one, which is related to the eye lens adjusting to the distance, is especially important in the application of HDMs being near-eye displays. Other limitations of the wide application of XR in medical practice are limitations of the HDMs themselves with their availability constraints limiting scalability and financial outlays related to their purchase [46]. Further sustainable development of XR technology is needed, balancing its technical parameters with costs, with a special emphasis on experimental validation [47].

Moreover, to provide a realistic user experience occlusion handling, especially in AR-based solutions, and proper rendering of objects are key issues [48]. A depth-based approach could extend the perception by capturing table depth data in real-time, for example, the presentation of the industrial scene in the form of a sparse point cloud, and conversion to the depth image [49]. Also, the development of 5G/6G edge computing and cloud servers contributes to the improvement of XR technologies [50].

An important research element is connected with human cognitions, namely explaining the relationship between user perception and the use of XR technology [51] by adjusting and analyzing users' personal experiences that are achieved through an awareness of the presented context. The challenge is being able to control the user's perception of XR relative to their perception of the real world [52].

7. Future plans

In the proposed solution, still, some technical issues connected with the quality of visualization, reduction of latencies, lighting defects, and orientation may be improved. In the future, we plan to evaluate the usefulness of the proposed system in surgical practice taking into account depth perception, quality of rendering, and visibility of anatomical landmarks, text, and objects, with the combination of the System Usability Scale (SUS) with the nontechnical skills (NOTECHS) rating scale for surgical teams [53, 54], and Mayo High-Performance Teamwork Scale (MHPTS) [55], which have been proposed for the evaluation of virtual surgical simulations in terms of an individual patient sample. The application of the Surg-TLX and others to the evaluation of human cognitions is also considered [56, 57].

Thus, the development of XR technology in medicine should be coordinated with the needs of healthcare in terms of security, precision, and reliability, taking into account the protection of personal data (sensitive information) as well as the development of XR-based devices. Improved XR-based technologies are also needed to enhance 3D Visualization. Because the quality and resolution of 3D imaging technologies will likely improve, providing even more detailed and accurate representations of the surgical field. This could include advancements in imaging modalities, such as higher-resolution CT image scans or real-time 3D ultrasounds, enabling surgeons to visualize anatomical structures with exceptional clarity.

Another important aspect is to improve connectivity and communication infrastructure and telecommunication technologies, which contribute to seamless and real-time communication between surgeons, remote experts, and other healthcare professionals. High-speed connections and low-latency transmission are crucial for transmitting large amounts of data, including high-resolution 3D images and video feeds, enabling efficient remote support.

In the future is also be the development and implementation of XR-based technologies and artificial intelligence-based algorithms as a part of the integrated system. It can be added value as an aid in surgical planning, real-time image analysis, and decision support. AI-based tools could help surgeons navigate complex anatomical structures, identify anomalies or critical regions, and provide insights based on vast amounts of medical data. Integrating AI into the 3D operating room with remote support could enhance surgical precision and outcomes.

Acknowledgements

This study was supported by the National Centre for Research and Development (Grant Lider No. LIDER/17/0064/L-11/19/NCBR/ 2020).

Conflict of interest

The authors declare no conflict of interest.

Notes/thanks/other declarations

This study was approved by the Medical Ethical Committee of the Jagiellonian University Medical College in Cracow, Poland No: 1072.6120.27.2020 and 1072.6120.92.2022.

Author details

Klaudia Proniewska[1,2], Damian Dolega-Dolegowski[3], Radek Kolecki[4,5], Magdalena Osial[6] and Agnieszka Pregowska[6]*

1 Jagiellonian University Medical College, Center for Digital Medicine and Robotics, Cracow, Poland

2 Department of Bioinformatics and Telemedicine, Jagiellonian University Medical College, Cracow, Poland

3 Jagiellonian University Medical College, Project LIDER XI NCBiR-HoloMed, Cracow, Poland

4 Department of Orthopedic Surgery, University Hospital in Cracow, Cracow, Poland

5 Jagiellonian University Medical College, Cracow, Poland

6 Institute of Fundamental Technological Research, Polish Academy of Sciences, Warsaw, Poland

*Address all correspondence to: aprego@ippt.pan.pl

References

[1] Kwok AOJ, Koh SGM. COVID-19 and extended reality (XR). Current Issues in Tourism. 2021;**24**:1935-1940. DOI: 10.1080/13683500.2020.1798896

[2] Xi N, Chen J, Gama F, Riar M, Hamari J. The challenges of entering the metaverse: An experiment on the effect of extended reality on workload. Information Systems Frontiers. 2023;**25**:659-680. DOI: 10.1007/s10796-022-10244-x

[3] Fang W, Chen L, Zhang T, Chen C, Teng Z, Wang L. Head-mounted display augmented reality in manufacturing: A systematic review. Robotics and Computer-Integrated Manufacturing. 2023;**83**:102567. DOI: 10.1016/j.rcim.2023.102567

[4] Zhang J, Lu V, Khanduja V. The impact of extended reality on surgery: A scoping review. International Orthopaedics. 2023;**47**:611-621. DOI: 10.1007/s00264-022-05663-z

[5] Dey A, Billinghurst M, Lindeman RW, Swan JE 2nd. A systematic review of 10 years of augmented reality usability studies: 2005 to 2014. Frontier in Robot AI. 2018;**5**:37. DOI: 10.3389/frobt.2018.00037

[6] Acidi B, Ghallab M, Cotin S, Vibert E, Golse N. Augmented reality in liver surgery. Journal of Visceral Surgery. 2023;**160**:118-126. DOI: 10.1016/j.jviscsurg.2023.01.008

[7] Wang J, Suenaga H, Liao H, Hoshi K, Yang L, Kobayashi E, et al. Real-time computer-generated integral imaging and 3D image calibration for augmented reality surgical navigation. Computerized Medical Imaging and Graphics. 2015;**40**:147-159. DOI: 10.1016/j.compmedimag.2014.11.003

[8] Malhotra S, Halabi O, Dakua SP, Padhan J, Paul S, Palliyali W. Augmented reality in surgical navigation: A review of evaluation and validation metrics. Applied Sciences. 2023;**13**:1629

[9] Gregory TM, Gregory J, Sledge J, Allard R, Mir O. Surgery guided by mixed reality: Presentation of a proof of concept. Acta Orthopaedica. 2018;**89**:480-483. DOI: 10.1080/17453674.2018.1506974

[10] Lungu AJ, Swinkels W, Claesen L, Tu P, Egger J, Chen X. A review on the applications of virtual reality, augmented reality and mixed reality in surgical simulation: An extension to different kinds of surgery. Expert Review of Medical Devices. 2021;**18**:47-62. DOI: 10.1080/17434440.2021.1860750

[11] Alotaibi YK, Federico F. The impact of health information technology on patient safety. Saudi Medical Journal. 2017;**38**:1173-1180. DOI: 10.15537/smj.2017.12.20631

[12] Lex JR, Koucheki R, Toor J, Backstein DJ. Clinical applications of augmented reality in orthopaedic surgery: A comprehensive narrative review. International Orthopaedics. 2023;**47**:375-391. DOI: 10.1007/s00264-022-05507-w

[13] Flood LM. Complications in otolaryngology–head and neck surgery. In: Bernal-Sprekelsen M, Carrau RL, Dazert S, Dornhoffer JL, Peretti G, Tewfik MA, editors. Thieme. Vol. 127. 2013. pp. 1043-1043. DOI: 10.1017/S0022215113002211

[14] Janssen BV, Theijse R, van Roessel S, de Ruiter R, Berkel A, Huiskens J, et al. Artificial intelligence-based segmentation

of residual tumor in histopathology of pancreatic cancer after neoadjuvant treatment. Cancers. 2021;**13**:5089

[15] Li X, Guo Y, Jiang F, Xu L, Shen F, Jin Z, et al. Multi-task refined boundary-supervision U-Net (MRBSU-Net) for gastrointestinal stromal tumor segmentation in endoscopic ultrasound (EUS) images. IEEE Access. 2020;**8**:5805-5816. DOI: 10.1109/ACCESS.2019.2963472

[16] Farronato G, Galbiati G, Esposito L, Mortellaro C, Zanoni F, Maspero C. Three-dimensional virtual treatment planning: Presurgical evaluation. The Journal of Craniofacial Surgery. 2018;**29**:e433-e437. DOI: 10.1097/scs.0000000000004455

[17] Mespreuve M, Waked K, Collard B, De Ranter J, Vanneste F, Hendrickx B. The usefulness of magnetic resonance angiography to analyze the variable arterial facial anatomy in an effort to reduce filler-associated blindness: Anatomical study and visualization through an augmented reality application. Aesthetic Surgery Journal Open Forum. 2021;**3**:ojab018. DOI: 10.1093/asjof/ojab018

[18] Bakhuis W, Sadeghi AH, Moes I, Maat APWM, Siregar S, Bogers AJJC, et al. Essential surgical plan modifications after virtual reality planning in 50 consecutive Segmentectomies. The Annals of Thoracic Surgery. 2023;**115**:1247-1255. DOI: 10.1016/j.athoracsur.2022.08.037

[19] Pratt P, Ives M, Lawton G, Simmons J, Radev N, Spyropoulou L, et al. Through the HoloLens™ looking glass: Augmented reality for extremity reconstruction surgery using 3D vascular models with perforating vessels. European Radiological Experiment. 2018;**2**:2. DOI: 10.1186/s41747-017-0033-2

[20] Zhang Y, Lu Y, Li J, Huang B, He X, Xiao R. Exploring the clinical benefits of mixed-reality technology for breast lumpectomy. Mathematical Problems in Engineering. 2023;**2023**:2919259. DOI: 10.1155/2023/2919259

[21] Ochs BG, Gonser C, Shiozawa T, Badke A, Weise K, Rolauffs B, et al. Computer-assisted periacetabular screw placement: Comparison of different fluoroscopy-based navigation procedures with conventional technique. Injury. 2010;**41**:1297-1305. DOI: 10.1016/j.injury.2010.07.502

[22] Klopfer T, Notheisen T, Baumgartner H, Schneidmueller D, Giordmaina R, Histing T, et al. Next step trauma and orthopaedic surgery: Integration of augmented reality for reduction and nail implantation of tibial fractures. International Orthopaedics. 2023;**47**:495-501. DOI: 10.1007/s00264-022-05619-3

[23] Proniewska K, Khokhar AA, Dudek D. Advanced imaging in interventional cardiology: Mixed reality to optimize preprocedural planning and intraprocedural monitoring. Kardiologia Polska. 2021;**79**:331-335. DOI: 10.33963/kp.15814

[24] Tagaytayan R, Kelemen A, Sik-Lanyi C. Augmented reality in neurosurgery. Archives of Medical Science. 2018;**14**:572-578. DOI: 10.5114/aoms.2016.58690

[25] Bennett KM, Griffith A, Sasanelli F, Park I, Talbot S. Augmented reality navigation can achieve accurate coronal component alignment during total knee arthroplasty. Cureus. 2023;**15**:e34607. DOI: 10.7759/cureus.34607

[26] Mongen MA, Willems PWA. Current accuracy of surface matching compared to adhesive markers in patient-to-image

registration. Acta Neurochirurgica. 2019;**161**:865-870. DOI: 10.1007/s00701-019-03867-8

[27] Nguyen NQ, Cardinell J, Ramjist JM, Lai P, Dobashi Y, Guha D, et al. An augmented reality system characterization of placement accuracy in neurosurgery. Journal of Clinical Neuroscience Official Journal of the Neurosurgical Society of Australasia. 2020;**72**:392-396. DOI: 10.1016/j.jocn.2019.12.014

[28] Qi Z, Li Y, Xu X, Zhang J, Li F, Gan Z, et al. Holographic mixed-reality neuronavigation with a head-mounted device: Technical feasibility and clinical application. Neurosurgical Focus. 2021;**51**:E22. DOI: 10.3171/2021.5.Focus21175

[29] Low D, Lee CK, Dip LLT, Ng WH, Ang BT, Ng I. Augmented reality neurosurgical planning and navigation for surgical excision of parasagittal, falcine and convexity meningiomas. British Journal of Neurosurgery. 2010;**24**:69-74. DOI: 10.3109/02688690903506093

[30] Jain S, Gao Y, Yeo TT, Ngiam KY. Use of mixed reality in neuro-oncology: A single centre experience. Life. 2023;**13**:398

[31] Ashrafian H, Clancy O, Grover V, Darzi A. The evolution of robotic surgery: Surgical and anaesthetic aspects. British Journal of Anaesthesia. 2017;**119**:i72-i84. DOI: 10.1093/bja/aex383

[32] Qian L, Wu JY, DiMaio SP, Navab N, Kazanzides P. A review of augmented reality in robotic-assisted surgery. IEEE Transactions on Medical Robotics and Bionics. 2020;**2**:1-16. DOI: 10.1109/TMRB.2019.2957061

[33] Chen Z, Marzullo A, Alberti D, Lievore E, Fontana M, De Cobelli O, et al. FRSR: Framework for real-time scene reconstruction in robot-assisted minimally invasive surgery. Computers in Biology and Medicine. 2023;**163**:107121. DOI: 10.1016/j.compbiomed.2023.107121

[34] Tanzi L, Piazzolla P, Porpiglia F, Vezzetti E. Real-time deep learning semantic segmentation during intra-operative surgery for 3D augmented reality assistance. International Journal of Computer Assisted Radiology and Surgery. 2021;**16**:1435-1445. DOI: 10.1007/s11548-021-02432-y

[35] Pregowska A, Osial M, Dolega-Dolegowski D, Kolecki R, Proniewska K. Information and communication technologies combined with mixed reality as supporting tools in medical education. Electronics. 2022;**11**:3778

[36] Kolecki R, Pręgowska A, Dąbrowa J, Skuciński J, Pulanecki T, Walecki P, et al. Assessment of the utility of mixed reality in medical education. Translational Research in Anatomy. 2022

[37] Peterson DC, Mlynarczyk GS. Analysis of traditional versus three-dimensional augmented curriculum on anatomical learning outcome measures. Anatomical Sciences Education. 2016;**9**:529-536. DOI: 10.1002/ase.1612

[38] Burian BK, Ebnali M, Robertson JM, Musson D, Pozner CN, Doyle T, et al. Using extended reality (XR) for medical training and real-time clinical support during deep space missions. Applied Ergonomics. 2023;**106**:103902. DOI: 10.1016/j.apergo.2022.103902

[39] Gan W, Mok TN, Chen J, She G, Zha Z, Wang H, et al. Researching the application of virtual reality in medical education: One-year follow-up of a randomized trial. BMC Medical

Education. 2023;**23**:3. DOI: 10.1186/s12909-022-03992-6

[40] Kumar A, Srinivasan B, Saudagar AKJ, AlTameem A, Alkhathami M, Alsamani B, et al. Next-gen Mulsemedia: Virtual reality haptic simulator’s impact on medical practitioner for higher education institutions. Electronics. 2023;**12**:356

[41] Leclercq WK, Keulers BJ, Scheltinga MR, Spauwen PH, van der Wilt GJ. A review of surgical informed consent: Past, present, and future. A quest to help patients make better decisions. World Journal of Surgery. 2010;**34**:1406-1415. DOI: 10.1007/s00268-010-0542-0

[42] Kyaw BM, Saxena N, Posadzki P, Vseteckova J, Nikolaou CK, George PP, et al. Virtual reality for health professions education: Systematic review and Meta-analysis by the digital health education collaboration. Journal of Medical Internet Research. 2019;**21**:e12959. DOI: 10.2196/12959

[43] Wilkat M, Karnatz N, Schrader F, Schorn L, Lommen J, Parviz A, et al. Usage of object matching algorithms combined with mixed reality for enhanced decision making in orbital reconstruction—a technical note. Journal of Personalized Medicine. 2023;**13**:922

[44] Eijlers R, Dierckx B, Staals LM, Berghmans JM, van der Schroeff MP, Strabbing EM, et al. Virtual reality exposure before elective day care surgery to reduce anxiety and pain in children: A randomised controlled trial. European Journal of Anaesthesiology. 2019;**36**:728-737. DOI: 10.1097/eja.0000000000001059

[45] Lee HJ, Drag LL, Bieliauskas LA, Langenecker SA, Graver CJ, O'Neill J, et al. Results from the cognitive changes and retirement among senior surgeons self-report survey. Journal of the American College of Surgeons. 2009;**209**(5):668-671.e662

[46] Labovitz J, Hubbard C. The use of virtual reality in podiatric medical education. Clinics in Podiatric Medicine and Surgery. 2020;**37**:409-420. DOI: 10.1016/j.cpm.2019.12.008

[47] Daher M, Ghanimeh J, Otayek J, Ghoul A, Bizdikian A-J, El Abiad R. Augmented reality and shoulder replacement: A state of the art review article. JSES Reviews, Reports, and Techniques. 2023. DOI: 10.1016/j.xrrt.2023.01.008

[48] Fang W, Hong J. Bare-hand gesture occlusion-aware interactive augmented reality assembly. Journal of Manufacturing Systems. 2022;**65**:169-179. DOI: 10.1016/j.jmsy.2022.09.009

[49] Li W, Wang J, Liu M, Zhao S. Real-time occlusion handling for augmented reality assistance assembly systems with monocular images. Journal of Manufacturing Systems. 2022;**62**:561-574. DOI: 10.1016/j.jmsy.2022.01.012

[50] Maddikunta PKR, Pham Q-V, Prabadevi B, Deepa N, Dev K, Gadekallu TR, et al. Industry 5.0: A survey on enabling technologies and potential applications. Journal of Industrial Information Integration. 2021;**26**:100257

[51] Wang Z, Bai X, Zhang S, Billinghurst M, He W, Wang Y, et al. The role of user-centered AR instruction in improving novice spatial cognition in a high-precision procedural task. Advanced Engineering Informatics. 2021;**47**:101250. DOI: 10.1016/j.aei.2021.101250

[52] Bernard F, Bijlenga PP. Defining anatomic roadmaps for neurosurgery

with mixed and augmented reality. World Neurosurgery. 2022;**157**:233-234. DOI: 10.1016/j.wneu.2021.09.125

[53] Brooke JB. SUS: A 'Quick and Dirty' usability scale. Paper/Poster presented. 1996

[54] Sevdalis N, Davis R, Koutantji M, Undre S, Darzi A, Vincent CA. Reliability of a revised NOTECHS scale for use in surgical teams. American Journal of Surgery. 2008;**196**:184-190. DOI: 10.1016/j.amjsurg.2007.08.070

[55] Malec JF, Brown AW, Leibson CL, Flaada JT, Mandrekar JN, Diehl NN, et al. The mayo classification system for traumatic brain injury severity. Journal of Neurotrauma. 2007;**24**:1417-1424. DOI: 10.1089/neu.2006.0245

[56] Fischer M, Fuerst B, Lee SC, Fotouhi J, Habert S, Weidert S, et al. Preclinical usability study of multiple augmented reality concepts for K-wire placement. International Journal of Computer Assisted Radiology and Surgery. 2016;**11**:1007-1014. DOI: 10.1007/s11548-016-1363-x

[57] Brown EJ, Fujimoto K, Blumenkopf B, Kim AS, Kontson KL, Benz HL. Usability assessments for augmented reality head-mounted displays in open surgery and interventional procedures: A systematic review. Multimodal Technologies and Interaction. 2023;**7**:49

Chapter 11

Perspective Chapter: Using Augmented Reality (AR) in the Education of Medical Bioengineers

Călin Corciovă, Robert Fuior, Andra Băeșu and Cătălina Luca

Abstract

Augmented reality (AR) is a technology that combines the real world with virtual elements, providing users with an enhanced interactive experience. AR has been used in a variety of fields, including medicine and bioengineering. In terms of training medical bioengineers, augmented reality can play a significant role in improving the learning process and understanding of human anatomy, medical procedures, and medical devices. Using AR technology, medical bioengineers can benefit from the following advantages in their training: three-dimensional visualization and interaction, medical procedure simulation, real-time guidance, collaboration and communication, medical device innovation and development. Using these technologies in the training of medical bioengineers, they can practice and become familiar with performing these procedures in a safe and controlled virtual environment. This can help increase confidence and practical skills before working in real life.

Keywords: augmented reality, medical bioengineering education, clinical immersion, simulation, medical devices, inter-professional communication

1. Introduction

These technologies provide an immersive and interactive digital scene for three-dimensional (3D) viewing medium, resulting in their widespread adoption in various fields that include commercial, educational, and biomedical sectors. Although the concept of virtual reality (VR) has existed since the nineteenth century, VR became popular during the 1990s. Technological advances in headsets and computer hardware, including computer graphics, led to many companies, especially in the entertainment sector, investing in this technology. VR headsets differ in platform, content, depth perception, tracking capabilities, viewing resolution, and audio technology. These devices significantly improve fields of view (FOV) and real-time frame rates that mitigate the effects of cybersickness to some extent [1].

Apart from VR devices, the augmented reality (XR) experience has also been on the rise in augmented reality (AR)/mixed reality (MR) devices. However, unlike VR devices, AR eyes are not widely commercialized due to their high cost. Despite the popularity of XR devices, a comprehensive analysis of the biomedical implications of this XR in medicine, surgery, and medical education is warranted. This overview defines the concepts of VR, AR, and MR and the capabilities of these technologies. First, current biomedical trends including visualization, clinical care, and research are summarized in XR, then VR and AR are used in the classroom as interactive teaching platforms. VR offers a fully virtual and immersive experience, while AR augments the real-world view with virtual information. MR performs real-time spatial mapping between real and virtual worlds. Interactions refer to the types of interactions that are enabled using technology.

VR allows interaction with virtual objects and AR allows interaction with physical objects. MR enables interaction between physical and virtual objects. Information refers to the types of data processed during display. In VR, the virtual objects displayed are registered in a 3D virtual space. AR provides real-time virtual annotation in the user's environment. In MR, viewed virtual objects are registered in 3D space and time in relation to the user's real environment [2]. All highly immersive XR experiences are presumably based on the seamless interaction of the physical and digital worlds. Therefore, the user's context, including what is in the user's surroundings and physical world, is particularly important. The contextual foundation of current AR applications well reflects this importance. Two specific examples are location-based and marker-based triggers for AR experiences [3] (**Figure 1**).

This helps determine a reference coordinate system in which virtual objects are located and tracked. Markers in AR experience ads may be visually replaced by other content, requiring segmentation. Machine learning-based approaches and classical

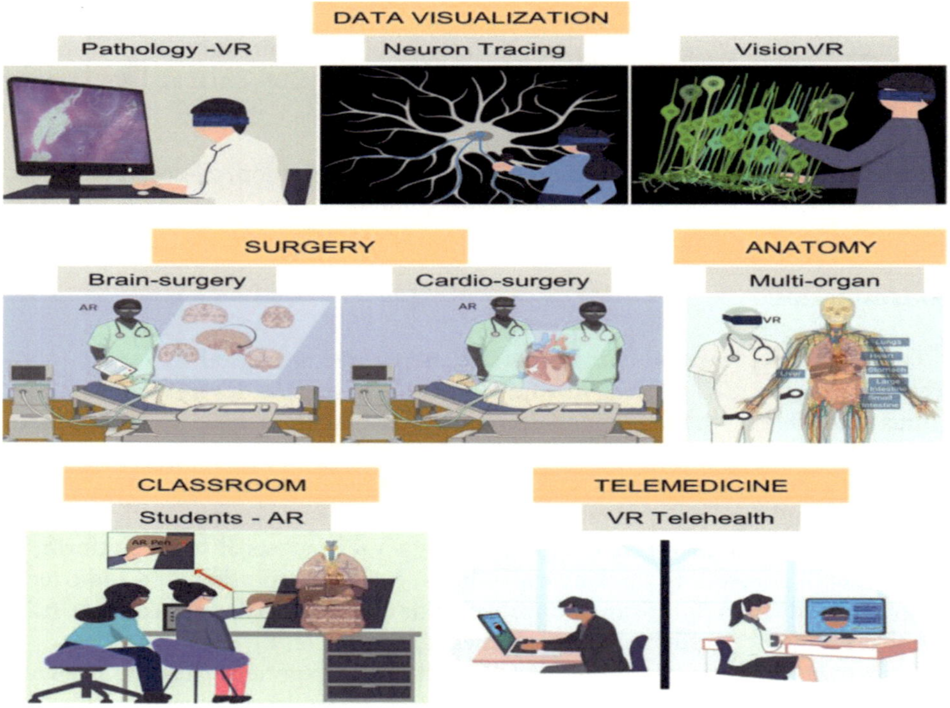

Figure 1.
VR- and AR-based visualization of scientific experimental imaging data, tools for surgery and anatomy, and collaborative interfaces for education and telehealth.

computer vision approaches are suitable for this. In many applications, markers are not replaced, but supplemented by additional information overlaid on the scene. For example, in training an auto mechanic, AR was used to label many of the components that the mechanic needed to identify within the complex engine assembly under the hood of a car [4].

2. Principles of augmented reality (AR)

The principles of augmented reality (AR) refer to the technological foundation and key concepts behind these innovative technologies. AR and VR enhance students' learning experiences by teaching biology, history, and geography concepts in interactive and engaging ways. For the generation leading the digital lifestyle, the use of media technology has significantly reduced our attention span. VR/AR as an educational tool offers a viable digital solution to this problem as it greatly reduces distractions and allows students to focus on the virtual space. One approach to using VR in the classroom is to provide students with headsets that synchronize with a central device and experience the same content. It can also be decentralized, where the lecture takes place in a virtual classroom and the student puts on her VR headset and connects from distinct locations. In addition to universities and medical schools, several K-12 classrooms have already introduced learning with XR technology [5]. For example, in a biology class, students can use their 3D model in VR to learn about the anatomy of the human body and other living organisms. At Agawam Public Schools in Massachusetts, teachers are incorporating Google Expeditions VR software into their classrooms to explore atoms and the inside of the human body. The software only requires a compatible smartphone and a Google Cardboard device. Through the expedition, students learned about history by experiencing the past and visiting ancient sites while sitting at their desks [6].

VR was used to teach cell biology concepts, impacting student participation and conceptual understanding. Students participate in the VR experience "Journey Inside a Cell" on The Body VR using an HMD and are tested against footage in a timed challenge to match each part of the cell to the correct label. Participants were also asked to complete a questionnaire describing their VR experience and whether it affected their learning. Of the students who participated in the study, 93.55% reported that VR enhanced their learning experience of cell biology concepts [7].

This overview of design principles will focus on specific strategies that instructional designers can use to develop AR learning experiences. These principles are contextualized in specific games or AR experiences developed by the Radford Outdoor Augmented Reality (ROAR) project at Radford University.

These two forms of AR (i.e., location-based, and vision-based) use multiple smartphone capabilities to create "immersive" and context-sensitive learning experiences in the physical environment, providing instructional designers with a new and potentially transformative tool for teaching and learning [8].

- *Information overlay*: One of the basic principles of AR is the overlay of virtual information over the real world. This principle involves the design and integration of computer-generated visual, audio, or haptic information over what the user sees or experiences in the real world through display devices, such as AR glasses. Graphics, images, or texts can be integrated and projected into the user's field of vision. This allows the combination of virtual elements with the physical environment in which the user is located [9].

The process of overlaying information in AR involves the following steps:

1. Capturing the real environment: To overlay information, AR needs a way to get information about the real environment the user is in. This is done using sensors such as video cameras, depth sensors, or motion sensors, which capture data about the object, surfaces, physical environment.

2. Identification of reference points: After the data about the real environment has been captured, the RA must identify the reference points in this environment. These reference points can be objects, surfaces, or special markers that are recognized and tracked by the system.

3. Calculation of the user's position and orientation: Based on the captured data and the identified reference points, the RA calculates the user's position and orientation in space. This information is essential to design and correctly place virtual elements in relation to the real environment based on tracking and geometric calculation algorithms.

4. Virtual information overlay: Once the position and orientation of the users are calculated, RA overlays the virtual information against the real average problem.

5. Interaction with overlay information: After the virtual information is overlayed, the user can interact with it. The interaction can be achieved through gestures, movements, voices, or the form of interaction with the AR device.

- *Real-time integration*: AR provides the ability to integrate real-time virtual information into the physical environment. This means that virtual elements can synchronize with the real moving environment and respond to user actions in a timely manner. Therefore, users can interact with virtual objects and observe real-time changes and feedback [10].

- *Position detection and tracking*: To correctly overlay virtual information on the real environment, AR practices position detection and tracking. This involves the use of sensors such as video cameras, depth sensors, or motion sensors to determine the user's position and orientation in space. This information is then used to correctly design and align the virtual elements [11].

- *Natural interaction*: Another important principle of AR is natural interaction with the virtual environment. Interaction technologies, such as voice recognition, gesture recognition, eye tracking or hand tracking, allow users to interact with virtual objects in an intuitive and natural way. This makes it easier to access and manipulate virtual information in an efficient and understandable way.

- *Spatial context*: AR places a strong emphasis on the user's spatial context. This includes understanding and recognizing the physical environment the user is in, so that virtual information can be integrated in a developed way. For example, by using object or marker recognition technologies, AR can identify and interpret objects and surfaces in the physical environment to project and place correct information [12].

3. Current biomedical trends in augmented reality (AR)

New widely used three-dimensional visualization technologies are virtual reality (VR), augmented reality (AR), and mixed reality (MR). These modern technologies also bring with them new challenges such as their excessive cost or the impact on human health, not yet sufficiently studied [13]. The usefulness of AR systems in medicine depends on the training of the technician, doctors, and teachers involved. To achieve maximum benefits, AR systems must be implemented with significant care and accuracy [14].

A new tool used in the visual assessment and manipulation of anatomical structures of real patients in 3D is represented by Applications of Virtual and Augmented Reality in Biomedical Imaging [15]. Following the studies, it was concluded that the use of AR in the visualization of radiological images offers doctors the possibility of making correct interpretations and can be used including in surgical planning [16].

Assisted surgery is also a field that has adopted AR technologies. These systems are used in distinct types of surgery, where AR can be used as a display or model with a promising perspective. Surgeon training apps fall into three distinct categories: echocardiography training, laparoscopic surgery, and AR and VR training for neurosurgical procedures [17].

The use of AR technologies such as HMD-based AR systems, augmented optics, augmented windows, monitors and endoscopes and their specific applications in the medical field are currently being discussed. The sense of touch can be transmitted to the user with the help of haptic augmented reality environments that can improve the work process [18].

Rehabilitation medicine successfully uses AR systems by implementing hand and arm movement systems in a spatial AR environment. In this way, a total patient immersion is created by creating a virtual audio and visual experience. The system guides the patient's rehabilitation tasks involving elbow, shoulder, and wrist movements. The system has the advantage of photographing the movements in real time and recording the patient's progress [19].

A new direction is the development of AR platforms that can be used for education. Here, you can include systems used in surgical operations, in the training of medical bioengineers or in the optimization of the medical educational process.

AR applications in medical education include several types of platforms that train students in diverse ways. There are systems based on AR scenarios that can be used to learn key concepts in VR environments with Google Cardboard, such as viewing single-cell protein images using an HMD and surgical planning using AR and VR. The software and hardware challenges of AR in biomedicine will not allow their large-scale development at this time [20].

The effectiveness of AR and VR systems in the fields of medical anatomy and health sciences is intensively studied by comparing the training outcomes of medical students who used VR and AR systems to those who used mobile applications, assessing the level of effectiveness. The study divided the target group of 59 students into 3 groups with different learning modes—VR, tablet-based applications, and AR who were taught a lesson on skull anatomy. Then their knowledge was assessed by repeating the experiments with different lessons. The results of the study showed that the AR and VR systems had benefits, because they promoted an increased involvement of students in the study process [21].

4. Biomedical education tools for teaching

The trend in the matter of learning at the university level is represented by the desire of teaching staff to instill in their students the ability to interpret information and at the same time to learn with joy. This trend can be achieved today by including modern technologies in student training without the need to give up solid learning principles. The challenge of the society, we live in, is given by students with significant differences in training and study motivation. This problem can be solved by integrating modern technologies in the teaching process. Today's students were born in the 2000s and expect modern universities to provide adequate digital infrastructure for teaching and learning. Medical education must adapt to many new and different healthcare contexts, including digitized health systems and students of the digital generation in a hyperconnected world. The instructional design must adapt to the target learners and the available resources. While the use of technology was already widespread in medical education, the COVID-19 pandemic has accelerated the need for more flexible, personalized, and collaborative learning.

Moral factors are paramount in the decision to pursue medical education. For this reason, critics argue that online medical education cannot compare to the instant feedback and sense of community offered by face-to-face courses. That is why the digitization process in medical education will pose important challenges in building empathy in medical practice.

The digitized curriculum in medical education can be developed following the principles of:

- *Interactivity.* Educational technology should promote interactivity in all learning environments, therefore, at this point, active learning requires a student-centered approach.

- *Bidirectionality.* A peer-to-peer relationship allowing students to apply their knowledge to solving complex problems with continuous feedback.

- *Mixture.* Modern technologies should integrate with traditional methods. Online lectures, VP, and online games must integrate traditional lectures, bedside teaching, and group simulations into a comprehensive curriculum.

- *Transnationality.* Medical study programs should be transnational based on web platforms that allow international cooperation. In this way, a homogeneity of training programs in European countries could be obtained.

- *Actuality.* The didactic materials should be revised and updated from year to year, even if the courses are available on platforms that the students can access from their private space. Materials must be carefully checked for timeliness and continuously refreshed [22].

We conducted a study in the specialized literature and identified the main types of tools used in biomedical engineering education:

1. Medical educators can adopt existing *digital and multimedia teaching* aids because they are based on a wide range of digital technologies. In the educational process, it could include technologies related to practice-based learning such as

video-based lectures [23]. In this way, active learning in the laboratory could be stimulated, the active participation of students in the classroom could be facilitated and group work encouraged [24].

2. *A safe and effective learning environment* that is rapidly developing in medical education is learning through simulation. Through this teaching method, basic knowledge of medical sciences such as anatomy, pharmacology, and physiology can be formed, rapid familiarization with medical procedures is produced, and solid clinical skills are created during simulated scenarios. Another benefit of simulation is that it can lead to the reduction of medical errors and the qualitative increase of the medical act [25].

The first simulator in education was introduced in the late 1990s and consisted of a life-size pelvis-type manikin with which midwives could practice during childbirth. Simulators have evolved, and today we have a high-fi "patient" simulator that talks, breathes, blinks, and moves like a real patient. Examples of simulation include: SimMan as a training and examination tool, ventriloscope to assess clinical examination skills among medical students (simulates auscultatory findings) [26].

A better approach to learning is to use a high-fidelity simulator that can be used in conjunction with medical devices for certain tasks. Also, simulation provides an ideal tool for assessing theoretical knowledge and evaluating practical skills of studies. Studies show that the use of simulators in medical education increases students' interest in the study [27].

3. *Applications for smartphones and mobile devices* were included in the learning process, which offer students the possibility to perform several tasks simultaneously. These applications actively contribute to the instant refresh of knowledge about diagnosis, medical management, patient health information, medical calculations, accessibility to contemporary clinical literature, continuation of medical treatment. These systems have an important impact on education and error prevention. The main problems with these systems are the risks of malware, potential privacy violations, and erroneous information in searches [27].

4. Core disciplines in the pre-clinical years at medical universities can be taught using *WSLA workstations (Workstation Learning Activities)*. These workstations are a flexible and scalable tool for moving toward integrated curricula. WSLA can be applied to large groups of students in a variety of contexts or environments. A wide range of clinical cases can also be used [28].

5. Virtual learning environments can be built with the help of *virtual reality (VR) and* applied in the educational process can lead to the improvement of the user experience by convincing the human brain that it is in a different environment [29]. Virtual environments can also be applied in special education and are useful in distance education. Perfecting skills is the main advantage of using VR in the medical educational process. Thus, future professionals will know how to adapt to different patients in different environments. At this moment, virtual patients or training systems are used for various therapeutic procedures [30].

6. *Virtual field trips (VFT)* can be used as an activity by pre-selecting web pages based on pre-defined topics that can be transformed into a structured online

learning experience. These virtual trips increase students' enthusiasm for learning and support the development of an active collaborative relationship with the teachers. The big disadvantage of these excursion environments is the contact with the real environment, the high cost, and the reduction of learning opportunities. In many situations, human interaction is undermined, and the flexibility offered by classroom collaboration between teacher and student is missing [31].

7. *Augmented reality (AR)* offers a complex view because it superimposes a computer-generated image on users' view of the real world. In the educational process and progress, the application is useful, especially for dynamic anatomy in real time. It offers a total experience to the user and allows, for example, the visualization of blood flow structures and even the performance of invasive procedures. A special utility is the application of AR in anatomical radiology, where radiological images from CT or MRI can be superimposed on a body. This creates a direct view of the spatial anatomy for the learner. If AR is combined with haptic technologies, tactile feedback can help users appreciate the consistency of distinct types of anatomical tissues [32].

8. We must also recognize the *limitations of AR application,* especially in the educational process. In these situations, powerful microprocessors are needed to drive AR because the devices used must be a natural extension of the surgeon's senses. The main characteristics that these devices should meet are low weight, high mobility, meet ethical, and deontological norms and be stable over time from the point of view of operation. A special and much-discussed problem is related to the confidentiality and management of patients' medical data, being an unsafe environment even at this moment.

9. The *Internet* has changed the entire process of teaching and educating students. It reduces barriers to knowledge sharing and acquisition. Online teaching has the same purpose as traditional teaching, which is to make learning possible. It does not eliminate the lecturer but supports and enhances educational activities. Meanwhile, online learning can also encourage the student to become a more critical and creative thinker capable of solving problems more easily. Currently, educational technologies for the development of teaching and learning through the Internet are widespread and relatively easy to use. An example in the Biomedical field is the EVICAB European Virtual Campus for Biomedical Engineering Portal, which is easy to build and apply to any discipline [33].

10. The future of technology in education will surely be *cloud technology,* as this technology facilitates access to Internet applications and services that allow information of any nature to be stored, shared, and accessed on any Internet-connected device. Cloud technology is already used in the educational process as a medium for storing and sharing digital textbooks, lesson plans, and assignments. Through the cloud, the student can access the materials before the class and the class can only debate the proposed topic. This approach will save time for the professional and the student, and the situation in which the student does not do his homework will no longer occur. The active application of cloud-type technologies in the educational process is limited currently by the security of stored and transmitted data [34].

11. In medical training, the concept of *Gamification* was also introduced through the development of medical training platforms using gamified elements and games. The advantage of gamification is the fact that it maintains the active involvement of the student in class, increases their degree of involvement, and provides immediate feedback. The students also find it enjoyable. The big problem at this moment is the conceptualization of games and their adaptation to medical education [35].

12. *Artificial intelligence (AI)* is a modern concept through which machine learning systems are created and can be applied at the educational level by automating grading and offering personalized learning opportunities for the type of student. AI has a dual benefit in education because it can help the teacher understand a student's learning patterns. Because personal involvement and interaction between doctor and patient in medical education are significant, there is an acute reluctance to use AI in the medical educational process. Another major problem is represented by ethical issues that prevent the application of AI in medical education [36].

13. An important innovation in medical education is represented by *Problem-Based Learning (PBL),* as it is defined as a student-centered approach to educational learning. It gives students with the opportunity to carry out research, fostering a spirit of collaboration. The PBL technique involves brainstorming activities and thus integrates and retains theory and practice in the application of knowledge and skills. The main challenge when implementing PBL is the significant amount of time required from the teacher for this activity, which is approximately four times higher than for a regular activity. PBL led to the implementation of evidence-based medicine (EBM), considered a revolution compared to classical empirical medical practice [37].

5. Virtual training for bioengineers and medical devices

Virtual training for medical bioengineering and medical device use can be a highly effective and valuable approach to educating professionals in bioengineering and medical device development. This method uses virtual reality (VR) and augmented reality (AR) technologies to provide immersive, interactive, and immersive training experiences. Some benefits and applications of virtual training for bioengineering and medical devices are presented:

1. *Hands-on experience:* Virtual training allows bioengineers to gain hands-on experience with medical devices and equipment in a safe and controlled environment. They can practice using complex instruments, perform virtual surgeries, and interact with medical devices without the need for physical prototypes or risking patient safety.

2. *Realistic simulations:* VR and AR can create highly realistic simulations of medical procedures and scenarios. Bioengineers can learn how to handle medical devices in various situations, prepare for potential challenges, and improve their decision-making skills in a risk-free setting (**Figure 2**).

3. *Collaborative learning:* Virtual training platforms can enable collaboration among bioengineers from distinct locations. They can work together on projects, discuss ideas, and learn from each other's experiences, fostering a more dynamic and engaging learning environment [11, 12, 37].

4. *Continuous skill improvement:* Virtual training can offer personalized feedback and performance evaluations, allowing bioengineers to track their progress and identify areas for improvement. This iterative learning process promotes continuous skill development and competence.

5. *Access to rare or expensive equipment:* Virtual training can provide access to medical devices that might be rare or expensive to use in traditional training settings. This opens up opportunities for bioengineers to familiarize themselves with a wide range of devices and technologies (**Figure 3**).

6. *Adaptability and flexibility:* Virtual training can be tailored to suit the specific needs of bioengineers and medical device developers. Content can be updated

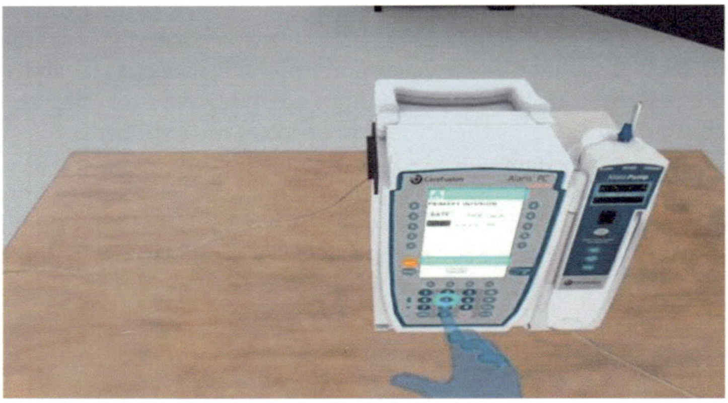

Figure 2.
Realistic simulation of medical device (infusion pump) [38].

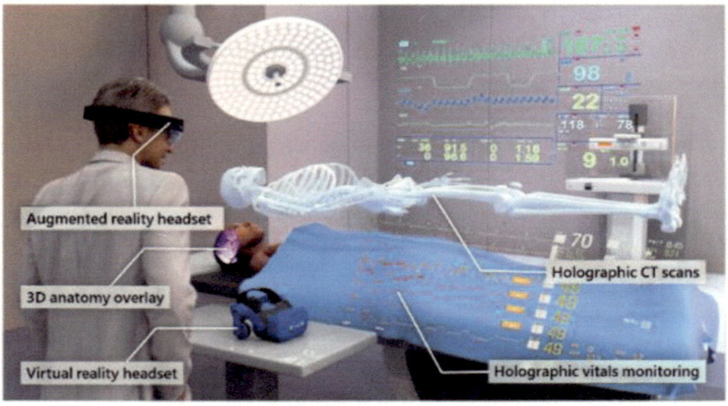

Figure 3.
AR in professional training of use CT equipment.

and adapted easily to align with the latest advances and regulatory requirements in the field.

7. *Remote training and telemedicine:* Virtual training can be delivered remotely, making it accessible to bioengineers and medical professionals worldwide. It also plays a crucial role in telemedicine, where medical device specialists can remotely assist healthcare providers in using complex equipment.

8. *Research and development:* Virtual training can be used during the research and development phase of medical devices. Engineers can simulate the performance of prototypes, conduct virtual tests, and identify design improvements before creating physical prototypes.

9. *Compliance and regulatory training:* Bioengineers and medical device developers need to be well-versed in regulatory requirements and quality standards. Virtual training can provide interactive modules on compliance and regulatory processes, ensuring adherence to industry guidelines.

Overall, virtual training offers a cost-effective, safe, and efficient way to enhance the skills and knowledge of bioengineers and medical device professionals. As technology continues to advance, virtual training is likely to become even more sophisticated and integral to the field of bioengineering and medical device development [39–41].

Calibration and maintenance of medical devices are critical processes to ensure their accuracy, reliability, and safety in healthcare settings. Augmented reality (AR) can play a significant role in simplifying and enhancing these tasks. Augmented reality (AR) can be applied to calibration and maintenance of medical devices.

5.1 Guided procedures

AR can provide step-by-step visual instructions overlaid onto the medical device, guiding technicians through the calibration and maintenance processes. This real-time guidance can help ensure that the procedures are performed correctly and consistently.

5.2 Interactive 3D models

AR can superimpose interactive 3D models of medical devices onto the physical devices. Technicians can manipulate the virtual components, disassemble the device virtually, and learn about its internal workings, helping in understanding the calibration and maintenance requirements.

5.3 Real-time data visualization

AR can display real-time data and diagnostics from the medical device directly in the technician's field of view. This allows them to monitor various parameters and make adjustments during calibration, ensuring the device is operating within the desired specifications.

5.4 Remote assistance

AR can facilitate remote assistance from experts during calibration and maintenance procedures. Technicians can wear AR-enabled glasses and collaborate with specialists who can provide guidance, review data, and offer solutions in real time.

5.5 Error detection and troubleshooting

AR can highlight potential issues or errors during the calibration and maintenance process. Technicians can quickly identify problem areas and take corrective actions, reducing downtime and minimizing errors (**Figure 4**).

5.6 Digital documentation and record keeping

AR can assist in documenting the calibration and maintenance processes digitally. The AR system can record each step and generate digital reports, ensuring proper documentation and traceability for compliance purposes.

5.7 Training and onboarding

AR can be used for training new technicians on how to calibrate and maintain medical devices. Interactive AR simulations allow trainees to practice these tasks in a virtual environment before performing them on actual devices.

5.8 Predictive maintenance

AR can integrate with the Internet of Things (IoT) sensors embedded in medical devices. By analyzing real-time data, AR can predict potential maintenance needs, allowing proactive servicing and preventing device failures.

Figure 4.
Error detection and troubleshooting simulated in AR [38].

5.9 Compliance and audit support

AR can provide access to calibration and maintenance records during audits, simplifying the process and ensuring adherence to regulatory requirements [42].

By leveraging AR in the calibration and maintenance of medical devices, healthcare facilities can streamline their operations, reduce downtime, enhance accuracy, and improve overall patient safety. However, it is essential to ensure that the AR systems used comply with relevant regulatory standards and maintain data security and privacy. As AR technology continues to evolve, its applications in healthcare are expected to become even more advanced and transformative.

Statistical analysis of the use of augmented reality in the education of medical bioengineers could provide a clearer picture of the impact of this technology on their learning. To perform such an analysis, data on the use of augmented reality in medical bioengineering education programs should be collected and a series of statistical analyses should be performed. Being a complex field that involves medical and engineering aspects, the study of the specialized literature returned only a few relevant works for the analyzed subject.

Some aspects that can be identified include the fact that the percentage of higher education institutions using augmented reality in their medical bioengineering programs remains low, mainly because the percentage of courses integrating augmented reality into the curriculum is limited. The assessment of initial and ongoing costs associated with the implementation of augmented reality in the education of medical bioengineers is quite high and would require a comparison of these costs with the benefits and improvements observed in the learning process. The long-term impact should be taken into account, including the monitoring of the careers of graduates who were trained with augmented reality during their studies and assessing their success in their respective fields. These statistical analyses could provide a comprehensive picture of the efficacy and impact of augmented reality in the education of medical bioengineers and could be used to make informed decisions regarding the further development and implementation of the technology in this field. It is important to note that the results of the analyses may vary depending on the specific implementation of augmented reality in a given program of study and the institutional context.

6. Conclusions

The use of augmented reality (AR) in the educational and professional training of medical bioengineers offers numerous significant advantages and transformative opportunities. Augmented reality provides a highly engaging and immersive learning experience, fostering better retention and comprehension of complex medical concepts and procedures. Bioengineers can interact with virtual models and simulations, enhancing their understanding and proficiency. AR enables medical bioengineers to practice in a safe and controlled environment, reducing the risk associated with real-life procedures. This allows them to build their skills and confidence before engaging in actual clinical settings.

AR systems can offer real-time feedback and performance assessment, allowing bioengineers to receive immediate guidance and correction. This iterative process improves learning outcomes and minimizes errors. AR encourages collaboration between medical professionals, engineers, and researchers. This interdisciplinary approach fosters innovation and leads to the development of more advanced and effective medical devices and solutions. The technology facilitates remote training and support, making it possible for bioengineers to access expertise and guidance from experts located elsewhere. This is particularly valuable in regions with limited access to specialized medical training.

AR enables continuous education and updates in the rapidly evolving medical field. Bioengineers can stay current with the latest medical advances and integrate them into their work. Implementing AR training solutions can be cost-effective in the long term. Once the initial investment is made, the technology can be reused for multiple training sessions, making it a valuable and accessible resource. The use of AR stimulates innovation and creativity among medical bioengineers. By exploring and experimenting with virtual models and scenarios, they can develop novel approaches and solutions to complex medical challenges.

Ultimately, the application of AR in the training of medical bioengineers translates to better patient care. Well-trained bioengineers can contribute to the development of safer, more efficient, and patient-friendly medical devices and technologies.

While AR offers substantial benefits, it is essential to recognize that its successful implementation requires careful planning, ongoing support, and proper integration into existing educational and professional training programs. As technology advances and becomes more accessible, the impact of augmented reality on the field of medical bioengineering is likely to continue growing, revolutionizing the way professionals are educated, and transforming healthcare for the better.

Acknowledgements

This work was supported by the project "Network of excellence in research and applied innovation for doctoral and postdoctoral study programs-InoHubDoc," code SMIS 153437.

Conflict of interest

The authors declare no conflict of interest.

Author details

Călin Corciovă[1*], Robert Fuior[1,2], Andra Băeşu[1] and Cătălina Luca[1]

1 Faculty of Medical Bioengineering, "Grigore T. Popa" University of Medicine and Pharmacy of Iasi, Romania

2 Faculty of Electrical, Energy and Applied Informatics Engineering, "Gheorghe Asachi" Technical University of Iasi, Romania

*Address all correspondence to: calin.corciova@bioinginerie.ro

References

[1] Kress BC, Cummings WJ. Optical architecture of HoloLens mixed reality headset. Digital Optical Technologies. 2017;**103350K**. DOI: 10.1117/12.2270017

[2] Rauschnabel PA, Rossmann A, tom Dieck MC. An adoption framework for mobile augmented reality games: The case of Poke´mon Go. Computers in Human Behavior. 2017;**76**:276-286

[3] Bin S, Masood S, Jung Y. Virtual and augmented reality in medicine. In: Feng DD, editor. Biomedical Information Technology. 2nd ed. London: Academic Press; 2020. pp. 673-686

[4] Khor WS, Baker B, Amin K, Chan A, Patel K, Wong J. Augmented and virtual reality in surgery-the digital surgical environment: Applications, limitations, and legal pitfalls. Annals of Translational Medicine. 2016;**4**:454

[5] Pottle J. Virtual reality and the transformation of medical education. Future Healthcare Journal. 2019;**6**:181-185

[6] Peugnet F, Dubois P, Rouland JF. Virtual reality versus conventional training in retinal photocoagulation: A first clinical assessment. Computer Aided Surgery. 1998;**3**:20-26

[7] Dash AK, Behera SK, Dogra DP, Roy PP. Designing of marker-based augmented reality learning environment for kids using convolutional neural network architecture. Displays. 2018;**55**:46-54

[8] Usher W, Klacansky P, Federer F, Bremer PT, Knoll A, Yarch J, et al. A virtual reality visualization tool for neuron tracing. IEEE Transactions on Visualization and Computer Graphics. 2018;**24**:994-1003

[9] Zimmerman E. K–12 Teachers Use Augmented and Virtual Reality Platforms to Teach Biology. 2019. Available from: https://edtechmagazine.com/k12/article/2019/03/k-12-teachers-use-augmented-and-virtual-reality-platformsteach-biology-perfcon

[10] Bennett JA, Saunders CP. A virtual tour of the cell: Impact of virtual reality on student learning and engagement in the STEM classroom. Journal of Microbiology & Biology Education. 2019;**20**(2):37

[11] Lavoie R, Main K, King C, King D. Virtual experience, real consequences: The potential negative emotional consequences of virtual reality gameplay. Virtual Reality. 2020;**25**:69-81

[12] Alalwan N et al. Challenges and prospects of virtual reality and augmented reality utilization among primary school teachers: A developing country perspective. Studies in Educational Evaluation. 2020;**66**:100876

[13] Venkatesan M, Mohan H, Ryan JR, Schurch CM, Nolan GP, Frakes DH, et al. Virtual and augmented reality for biomedical applications. 2021;**2**(7):100348. DOI: 10.1016/j.xcrm.2021.100348

[14] Patel D, Shah D, Shah M. The intertwine of brain and body: A quantitative analysis on how big data influences the system of sports. Annals of Data Science. 2020;7(1):1-16. DOI: 10.1007/s40745-019-00239-y

[15] Izard SG, Juanes Méndez JA, Palomera PR, García-Peñalvo FJ.

Applications of virtual and augmented reality in biomedical imaging. Journal of Medical Systems. 2019;**43**:102. DOI: 10.1007/s10916-019-1239-z

[16] McCoy CE, Sayegh J, Alrabah R, Yarris LM. Telesimulation: An innovative tool for health professions education. AEM Education and Training. 2017;**1**(2):132-136

[17] Barsom EZ, Graafland M, Schijven MP. Systematic review on the effectiveness of augmented reality applications in medical training. Surgical Endoscopy. 2016;**30**(10):4174-4183. DOI: 10.1007/s00464-016-4800-6

[18] Scharver C, Evenhouse R, Johnson A, Leigh J. Designing cranial implants in a haptic augmented reality environment. Communications of the ACM. 2004;**47**(8):32-38. DOI: 10.1145/1012037.1012059

[19] Hondori HM, Khademi M, Dodakian L, Cramer SC, Lopes CV. A spatial augmented reality rehab system for post-stroke hand rehabilitation. Studies in Health Technology and Informatics. 2013;**184**:279-285

[20] Liu Y, Zhang Y, Zhang L, Bai H, Wang G, Guo L. The impact of SimMan on resident training in emergency skills. Medicine (Baltimore). 2019;**98**(2):e13930

[21] Moro C, Štromberga Z, Raikos A, Stirling A. The effectiveness of virtual and augmented reality in health sciences and medical anatomy. Anatomical Sciences Education. 2017;**10**(6):549-559. DOI: 10.1002/ase.1696

[22] Filetti S, Saso L. Innovative Medical Education in the Digital Era. Roma: MCGraw Hill; 2022. pp. 6-16

[23] Dominguez M, Di Capua D, Leydon G, Loomis C, Longbrake EE, Schaefer SM, et al. A neurology clerkship curriculum using video-based lectures and just-in-time teaching (JiTT). MedEdPORTAL. 2018;**14**:10691

[24] Atlantis E, Cheema BS. Effect of audience response system technology on learning outcomes in health students and professionals: An updated systematic review. International Journal of Evidence-Based Healthcare. 2015;**13**(1):3-8

[25] Khan K, Pattison T, Sherwood M. Simulation in medical education. Medical Teacher. 2011;**33**(1):1-3

[26] Verma A, Bhatt H, Booton P, Kneebone R. The Ventriloscope® as an innovative tool for assessing clinical examination skills: Appraisal of a novel method of simulating auscultatory findings. Medical Teacher. 2011;**33**(7):e388-e396

[27] Qayumi K, Pachev G, Zheng B, Ziv A, Koval V, Badiei S, et al. Status of simulation in health care education: An international survey. Advances in Medical Education and Practice. 2014;**28**(5):457-467

[28] Latif MZ, Hussain I, Saeed R, Qureshi MA, Maqsood U. Use of smart phones and social media in medical education: Trends, advantages, challenges, and barriers. Acta Informatica Medica. 2019;**27**(2):133-138

[29] Riva G, Wiederhold BK, Mantovani F. Neuroscience of virtual reality: From virtual exposure to embodied medicine. Cyberpsychology, Behavior and Social Networking. 2019;**22**(1):82-96

[30] Baumann-Birkbeck L, Florentina F, Karatas O, Sun J, Tang T, Thaung V, et al. Appraising the role of the virtual

patient for therapeutics health education. Currents in Pharmacy Teaching and Learning. 2017;**9**(5):934-944

[31] Kim Y, Kim H, Kim YO. Virtual reality and augmented reality in plastic surgery: A review. Archives of Plastic Surgery. 2017;**44**(3):179-187

[32] Chittaro L, Ranon R. Web3D technologies in learning, education and training: Motivations, issues, opportunities. Computers in Education. 2017;**49**(1):3-18

[33] Malmivuo J. Innovative Teaching Practice and Assessment with Technology Applications in International Biomedical Engineering Education, A Volume in the Advances in Higher Education and Professional Development (AHEPD) Book Series Chapter. Vol. 2014. IGI-Global; 2014. pp. 123-148. DOI: 10.4018/978-1-4666-5011-4.ch011

[34] Liu W-L, Zhang K, Locatis C, Ackerman M. Cloud and traditional videoconferencing technology for telemedicine and distance learning. Telemedicine Journal and E-Health. 2015;**21**(5):422-426

[35] Gentry SV, Gauthier A, L'Estrade Ehrstrom B, Wortley D, Lilienthal A, Tudor Car L, et al. Serious gaming and gamification education in health professions: Systematic review. Journal of Medical Internet Research. 2019;**21**(3):e12994

[36] Challen R, Denny J, Pitt P, Gompels L, Edwards T, Krasimira Tsaneva-Atanasova K. Artificial intelligence, bias, and clinical safety. BMJ Quality & Safety. 2019;**28**(3):231-237

[37] Cirillo D, Valencia A. Big data analytics for personalized medicine. Current Opinion in Biotechnology. 2019;**58**:161-167

[38] NVRT Labs INC. Available from: www.nvrtlabs.com [Accessed: June, 2023]

[39] Wu H-K, Lee SW-Y, Chang H-Y, Liang J-C. Status, opportunities, and challenges of augmented reality in education. Computers in Education. 2013;**62**:41-48

[40] Green C, Harrington C. Student-Centered Learning: In Principle and in Practice. Lansing, MI: Michigan Virtual University; 2020

[41] Singh A, Ferry D, Mills S. Improving biomedical engineering education through continuity in adaptive, experiential, and interdisciplinary learning environments. ASME Journal of Biomechanical Engineering. 2018;**140**(8):0810091

[42] Singh A, Ferry D, Balasubramanian S. Efficacy of clinical simulation based training in biomedical engineering education. ASME Journal of Biomechanical Engineering. 2019;**141**(12):121011